The Last Neanderthal

THE LAST NEANDERTHAL

Understanding How Humans Die

Ludovic Slimak

Translated by Andrew Brown

polity

Originally published in French as *Le dernier Néandertalien. Comprendre comment meurent les hommes* © Odile Jacob, 2023

This English translation © Polity Press, 2025

This book is supported by the Institut français (Royaume-Uni) as part of the Burgess programme.

Polity Press
65 Bridge Street
Cambridge CB2 1UR, UK

Polity Press
111 River Street
Hoboken, NJ 07030, USA

ISBN-13: 978-1-5095-6958-8 – hardback

A catalogue record for this book is available from the British Library.

Library of Congress Control Number: 2025933326

Typeset in 11.5 on 14 Adobe Garamond
by Fakenham Prepress Solutions, Fakenham, Norfolk NR21 8NL
Printed and bound in the UK by CPI Group (UK) Ltd, Croydon

The publisher has used its best endeavours to ensure that the URLs for external websites referred to in this book are correct and active at the time of going to press. However, the publisher has no responsibility for the websites and can make no guarantee that a site will remain live or that the content is or will remain appropriate.

Every effort has been made to trace all copyright holders, but if any have been overlooked the publisher will be pleased to include any necessary credits in any subsequent reprint or edition.

For further information on Polity, visit our website: politybooks.com

Contents

First words

'But, Dad, it's already sad enough that butterflies die.'

My young son has just brought me a butterfly wing, torn to pieces like a small piece of fragile fabric. A fabric in tatters. He looks so moved that my words come out too quickly. 'It's not serious, sweetheart, he may have died a natural death, because he was too old …' As if the natural death of a butterfly could be acceptable for a small human just a handful of years old. My smile fills him with questions. As if I just didn't see it. As if I no longer knew how to see important things.

And yet, yes, it's *already* sad enough that butterflies die.

And this *already*, powerful, unavoidable, fact traces an invisible border. We have *already* entered into the lands of sadness.

That morning, my heart was heavy again, all because of a tattered wing. Was I sad about the butterfly, or because my young son had revealed to me that I no longer knew how to feel for the things that really matter? Perhaps we are always sad for ourselves when we lose the ability to see those invisible borders.

That same day, in the dust of a prehistoric cave, I was to discover the remains of one of the last Neanderthals. The vestiges of this body, of this distant death, were not just the sign of a man's death, like the remains of the butterfly. He had died as if behind him all the butterflies of his species had to die. He had died, and his kinsfolk too. We will all die one day. But, imagine all of us dying together. Like a cloud of dead people without children. Or in a slow death where, generation after generation, the army of children is outnumbered by the army of the dead.

Is this how humans die? Or do they die as they live, as Aragon and Ferré have taught us, *and their kisses follow them from afar*?[1] But what if even their kisses no longer followed them? What if all humans died as one? Died a natural death? Because they were too old?

No one knows how humans die. But die they do.

This book gathers together my thoughts on this kind of death, which is singular, unexpected, and yet inevitable. Death is always there, lurking, and so inevitable that perhaps it should not raise any questions for us. As if we should be surprised at surviving, not dying. As if the anomaly were not the death of all the humanities that once walked alongside us but rather the survival of our species alone. Death haunts our minds, always, as if we awaited it at every corner, like the arrival of a friend far too old to embrace. Will this friend come, at last? Or will humanity endure, unscathed, never to be cut down by the Old One with the scythe?

And, when I was suddenly confronted in August 2015 with the astonishing discovery of one of the last Neanderthals, I could think of nothing but the path this man took, before departing forever, with all his people.

This man, this dead man, this death, may have something to tell us. For our own path. Thinking of tomorrow, I wonder how humans die.

1

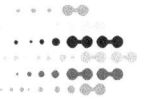

Unexpected complications on the way to the unthought

Boldly risking a step into the unknown

In old Neanderthal caves there's something uncertain about the smell of earth and flint. It is decaying, dry and damp at the same time, a bit like the smell of broken spring soil; a place where the living and the dead grow out of each other and can thus give voice.

Something almost unpleasant, but something that still makes you want to smell it again, to come back to it. As if to remember that scent, quite unlike anything else. Is it the scent of millennia? From all the cooking, the rotting, the infinite amount of stuff left behind, emitting this little whiff of sourness? All those bones, that bone dust, that humus of meat. Those thousands of generations of love; those thousands of generations of death.

I don't know. Maybe those smells aren't so old, when you think about it. Maybe they just indicate recent putrefaction, all those bacterial decompositions, those mosses, those algae and all the modest life of cave floors? There are things happening in these soils. There is nothing inert about them: they continue their small lives, quietly, at their microbial pace. These invisible lives are all around us.

But, for some time now, there is something else. It's no longer the familiar harsh little smell that invades my olfactory space when I dig around in the ancient Neanderthal soils.

Something else haunts these places.

Regularly, or rather just occasionally, powerful smells of grilling emerge from the soils I'm digging in. It's ... how can I put it? – like fire, burning, ash, coal. No. No. It's not really that. It's stronger, more intoxicating. Like … Like a gamey aroma. Yes, that's it, grilled flesh, charred game, burnt fat. Yes, a powerful smell of grilled game. The smell seems to emerge from the ground so strongly that, for a moment, the air is almost unbreathable. And the smells vary from place to place and with the depth of my digging. Here, there's something soft, almost subtle, about it: but a few days later, in a different area of the cave and in older soils, the smell of charred flesh is so strong that for short moments I stop digging. I breathe it in, trying to find where these incredible aromas come from. How on earth? We are on archaeological soils that are more than fifty-five millennia old. What's going on with these prehistoric odours? Neanderthal smells. It makes no sense. And yet, they're here, powerful, obvious, unavoidable – and I don't know what it means. No one has ever reported smells from the age-old depths of caves. But these smells suddenly permeate the entire space, without warning, and disappear just as quickly. But they're certainly here. I can't pretend otherwise. I can't deny them. Something has survived and still haunts these places. Something still breathes this life of yesteryear, and one's nose isn't mistaken. It seems insane, but we have to deal with it. That's how it is. Odours clearly linked to the activities of Neanderthal populations have transgressed time, inviting themselves to gather around us tens of millennia later.

But these conclusions, this awareness, became clear to me only in 2006. I'd already been searching tirelessly in caves, three to four months a year, for almost fifteen years. Fifteen years of smelling without understanding and, more importantly, without accepting the facts. In fact, I think my senses perceived it, that's clear enough; but the impossibility of the situation stopped me from becoming fully aware of an objective reality. That's the whole problem in science. We do science only when we take a step into the unknown. When we agree to consider a possibility that doesn't seem reasonable. When we consider a solution that instinctively seems a little ridiculous. But doing science isn't dressing up in a white coat and running computers to crack infinite equations. In the first instance, doing science is considering the improbable. The instinctively ridiculous. And taking this leap into the void. And the problem is the way we rely on a parachute. On all the unconscious presuppositions that muffle our thinking, bringing it delicately back down into our comfort zones. The leap into the unknown isn't just painful; it's a fight against ourselves, against the countless filters that frame us, supervise us, police us, prevent us from thinking in real freedom; prevent us for fifteen years from understanding that it smells like ... grilled game, for heaven's sake! That it smells *horribly* like game. The brain filters things, brings us back to the familiar. Always. Back to the possible. Back to the reasonable, the reassuring. I thought I was free. We always think we're free, in general. But we're always prisoners of ourselves. However, we have to take it, sometimes – this risky step into the unknown. Accepting this transgression should constitute the first lesson of scientific thought. Learning to free ourselves, to reject our logic. No longer to reject the impossible. To revise, at each step, the field of possibilities that frames us. To love doubt. Do you want to know what a true scientific discovery is? Well, it's understanding something that seems impossible to us. More precisely, it's the demonstration that something that seems a bit ridiculous, a bit laughable, is a reality. If the discovery doesn't rub up against common sense, doesn't chip away at it, you can be sure that the discovery is of very secondary importance. It doesn't constitute a step. A crossing. If you transgress the common notion of reality, if the idea seems so ridiculous that you'd hardly dare to put it in words, then you're touching on science. You've taken this step, this transgression towards the unknown. But, by definition, consciousness and common sense block information. They block any truly free thought. Don't you forget it!

And then comes the slap in the face. At a certain point, the unconscious establishes connections. Serial connections. Calculations. Equations of which we have absolutely no perception and which, at the least expected moment, hand over the answer to consciousness. We don't know why the answer arrives at this precise moment. And we don't know why it took so long to express itself.

I remember that moment very precisely.

I was knees in the dirt, as usual; I got up, I picked up my bucket of fifty-thousand-year-old sand and I left the cave. I suddenly had a conviction that was both improbable and overwhelming, like the tunes you hear in the morning that cling to you, turn into ear-worms, going endlessly round and round. You see? I had to check. I had to test this strange conviction, this nagging idea. I sifted through these prehistoric sands, as usual, to be sure I hadn't missed a bit of flint or a little piece of bone when digging through the archaeological soil. But, instead of throwing away the little stones, I kept them. I broke them open. I brought them up to my nose. I breathed them in. With each crack, with each rupture of the surfaces, an explosion of smells. The scents weren't in the soil, they were frozen in the stone. These stones are fragments of the ancient vault of the cavern. They have fossilized all the soot deposited on the walls when the Neanderthals lit their fires in the cave. And, what I have before my eyes, after breaking my stones, is a real barcode of blacks and whites: the black of the soot, the white of the calcite concretions that cover them and fossilize them. They're there, my smells, trapped in the stone, like an immaterial fossil. The fossil of past breaths.

It would take us more than ten years to get these black and white edges to speak but, from 2006 onwards, we had a gut feeling, a certainty, that we would be able to transgress time.[1]

Another place, another time?

It didn't look like much, of course, I hadn't just discovered the Sistine Chapel or Tutankhamun's tomb, true – but these little pebbles would be one of the keys to rethinking the encounter between Neanderthals and that damned *Sapiens*.

Because, on these pebbles, it wasn't just the smell that had been trapped. It was time.

When we archaeologists dig, we have no notion of time. No idea of the precise age of the flint object that we've just extracted from the earth. We measure, we date, using the formidable tool of carbon 14. Suddenly, we know that the last Neanderthal is forty-two thousand years old. We also know that the first *Sapiens* on this same territory is, likewise, forty-two thousand years old. But Neanderthal and *Sapiens* exist in time loops where the uncertainty factor is more or less one thousand years.

My little stones, my barcodes, would enable us to perceive real time, and no longer the time of physicists – they would mean we could, year after year, watch the seasons passing. The calcite gradually deposits itself on the walls, a thin layer in the dry season, a thin layer in the wet season and, for the first time there are human traces, trapped between these thin layers of stone. We could finally look at time and finally analyse Neanderthals in their temporal reality, in six-month steps. We'd probably never do any better than this astonishingly high degree of resolution, unless we invented the time machine. This wonderful scientific advance, fuliginochronology (from Latin *fuliginosus, fuligo*: soot), was to be accompanied, in the same cave, by the discovery of human teeth. Nine, no more than that. Nine teeth for thirty years of work. Each of these teeth required on average ten months of continuous digging in the cave, by about fifteen people. Each of these teeth required a lot of effort. The cave opens due north, facing the Mistral. And this wind often roars, howls in our ears. I remember it perfectly, that same year 2006: the wind never stopped. It started blowing one morning, without warning anyone, for no reason perceptible to the poor archaeologist crouching in the cave, with the winds rushing in. In that year, 2006, four weeks later, it was still blowing. The Mistral is a wind that drives you crazy. A cold, brutal wind. It rushes in, and freezes in your bones, again and again. After that month of painful toil I was exhausted, as if stunned. Yet it was this same wind that had deposited the sands in the cave, fossilizing the prehistoric settlements. In the Mandrin cave² we no longer knew if the Mistral was a curse or a blessing. Anyway, we paid dearly for those nine human teeth, which cost thirty years of digging. But what they would tell us was unexpected, and of a great scientific and historical power. These teeth are almost all baby teeth, lost by children of the Palaeolithic. The oldest were lost here more than one hundred millennia ago. The last ones were lost between forty and fifty thousand years ago at the time of the Neanderthal extinction. Nine small teeth for sixty long millennia. It's not much. But it's one of the most important continuous anthropological records discovered in a single place in ancient Europe. The analysis would show, unsurprisingly, that all these teeth were Neanderthal. But the fourth tooth, from the fifty-fourth millennium, is another story.

This tooth fundamentally changes what we thought we knew about one of the greatest turning points in the history of Europe. It reminds us that we know so little, and that our understanding of the very ancient story of the last Neanderthals and the first modern humans on our continent is based on very little evidence. This tooth from the fifty-fourth millennium is supposed to be Neanderthal. We know that *Sapiens* wouldn't reach the European continent until ten millennia later, right?

No.

No, this tooth tells us another story. Analysis reveals that it belonged to a child. A little thing of between three and six years old who died fifty-four millennia ago. Its morphologies, analysed in very high resolution, would reveal to Clément Zanolli, one of the CNRS[3] members of my teams at the Mandrin cave, that this child was not a Neanderthal but indeed an ancient, Palaeolithic *Homo sapiens*. There wasn't the slightest possible doubt. We're making a gigantic leap in time. No one expects to come across remains attributed to *Sapiens* in Europe in such remote periods. Again, if you'd told me it was in western Turkey, or in Greece, or in the depths of the Caucasus, those immense blocks of mountains between the Black Sea and the Caspian Sea that separate Europe from Asia, that would be understandable – but here, in the Rhône Valley, to the far west of Europe, it just doesn't fit. However, in February 2022, my teams and I would publish our conclusions about these nine small teeth. A discovery that would set the media ablaze across the world.

It's incomprehensible, unexpected, unthinkable! We're at the opposite end of the Mediterranean from the eastern shores where we know that *Sapiens* loafed around for at least one hundred and fifty millennia but without ever turning west, without ever looking towards Europe.

Unthinkable?

Beware of such conclusions and of all those people who might feel emboldened to make carping comments precisely because it all seems totally unthinkable. In reality, it was all considered, pondered, assessed, published – years before stating that this famous tooth belonged to *Homo sapiens*. But faced with the unthinkable, people sometimes prefer not to read, not to consider, not to quote. Five years before the publication announcing this small tooth of a modern human being from the fifty-fourth millennium, the comparative analysis of the thousands of flints found around this same small tooth had already enabled us to conclude that these technologies had to be attributed to *Sapiens* alone.[4] From then on, the demonstration no longer rested on the discovery of a small isolated tooth but on thousands of remains that clearly pointed to the very ancient presence of modern humans in the Mediterranean areas of Western Europe.

What we find, on the ground around the remains of this little *Homo sapiens*, are flint points, hundreds of points, probably thousands. We'd discovered 1,499 of them in the fifty square metres excavated in the cave, but the site is much larger, it extends well beyond the walls of the small cave, and it's perfectly preserved beyond its entrance, over much larger areas. These thousands of flint points litter the ground. They are magnificent, mass-produced, and all identical to within one or two millimetres, literally standardized. I'd never seen this in Neanderthals. No one had ever seen this in Neanderthals. I named these very special flints 'Neronian' in 2004, without having the slightest idea who could be the author of these remarkable technologies. Look, it's not because it's remarkably modern and standardized that a Neanderthal *couldn't* have been its author. The creature can amaze us, you know! Who knows, anyway? And this 'who knows?' has always been my position. A strong position. Of principle. We just don't know. But if Neanderthals really made it, they're like us. I mean they think like us. They understand the world like us. Although, in 2004, I hadn't drawn any definitive conclusions about the author of this Neronian, those last three sentences were deeply rooted in my mind. If Neanderthals can really make that, I can understand their crafts, immediately, at first glance. As experimenters, as flint knappers, as artisans, as humans. I can understand right away, easily, instinctively. So I was doubtful when I defined this Neronian. And on the definition of the Neronian, for me, everything turned. Neanderthals were either the same as us, or the astonishing creatures that all this handiwork throughout Europe enabled me to perceive. Everywhere I went, I hunted the creature. It was this quest and this creature that I described in *The Naked Neanderthal*.[5]

I had no ulterior motive. No pre-established thought. Doubt, as always – a doubt that overwhelms me, fills me, obsesses me. Is it Neanderthal? Is it *Sapien*? But I don't know, for goodness sake. Everything is possible. Why couldn't a Neanderthal do that? Why not? No one has the slightest idea.

I don't know.

French poems …

But in 2016 I was to be confronted by a new technological shock. We're in the United States, in Massachusetts, at Harvard University. In a sort of underground bunker where, after getting us to sign an endless list of administrative papers, the curators of the Peabody Museum authorize us to open the drawers containing collections from the Mediterranean East deposited here since the end of the Second World War. To do this, we'll have to dress up as astronauts, or caricatures of mad scientists, with white coats, hairnets, hairnets around our shoes, surgical gloves, masks, security badges with codes to be repeated twice or else all the sirens in the building will go off – a punishment that will be administered several times in the eight months of our stay in the underground bunker. It's always a bit embarrassing when we hastily call the curators to explain that we've put in a wrong number, or that we've taken too long to enter the code, or that we've got mixed up somewhere between the procedure for opening and closing the armoured door. Oh yes, damn, when we go out we have to enter the code *before*, not *after*, the badge is scanned! The curators roll their eyes, once again. These French, they're such poets! But how serious it is here – and what a lack of poetry in these underground passages with their metal cabinets. In these cabinets, there are flints from fifty thousand years ago, but in those other ones, four levels lower, there are bows, arrows, quivers, painted and decorated skins, telling the stories of the conflicts between the peoples of the plains and the European settlers.

Another place, another time. Another history? Or a hesitant utterance carried through the ages?

Wandering through the naves of our thermonuclear cathedrals

I've come here to open the drawers where objects from the fortieth to the fiftieth millennium lie. They contain tens of thousands of flints from an immense prehistoric shelter in the eastern Mediterranean, on the slopes of Mount Lebanon. Between 1937 and 1948, archaeological operations revealed traces of prehistoric settlements unique in Eurasia. Here, we can see other ways of being in the world emerging very gradually, millennium after millennium – the ways of the old *Sapiens* who, even fifty millennia ago, already clearly expressed our ways of conceiving our universe: and this conclusion alone already opens a thousand windows on what our humanity is, while questioning other ways of being in the world. How the old Neanderthal traditions die, and how they are swept away by these new ways of being in the world – by these new ways of thinking about the world.

How did these thousands of flints end up here, on the east coast of the United States? Probably a bit like those Egyptian bas-reliefs and

Mesopotamian frescoes, from the time when Westerners dug everywhere in search of the most beautiful *this*, the oldest *that*, in search of the relic that would make the difference. The one that would bring in the punters from hundreds or thousands of kilometres around, a guarantee of fame. A guarantee that the museum would be exceptional. A guarantee of prestige, a guarantee of renown. A guarantee that it could last for centuries and centuries …

The old cult of relics in the Middle Ages, which ensured that the bodies of the saints would be cut up and distributed throughout the most beautiful cathedrals of Europe, draws an invisible line connecting our museums to the huge naves of our ancient devotions. Not only did the dismembered bodies of sanctified humans able to pester the one God reside here, but also the remains of fabulous creatures leading to the doors of every kind of magic. You will find in the cathedral of Saint-Bertrand-de-Comminges, still on display, a dried crocodile suspended by its tail. The scaly beast is said to have haunted the Pyrenean forests until the bishop of Comminges defeated it in its lair. In the cathedral treasure you'll also find two huge unicorn horns. Here the bishops ventured to the gates of another world. And if you go to the Sainte-Chapelle in Paris, you'll be presented with a griffin's paw.

It must be acknowledged that the cult of relics no longer draws crowds in old Europe, but we've taken care to transfer to our museums the relics of modern times, the recomposed memories of all ages and all continents. Recomposed, but on artificial respiration. They are evocations, of course. Modern mythologies that only speak for a certain time, since knowledge evolves. Perspectives change. Museums are constantly recomposed in tune to changes in our knowledge and the sensitivities of our perspectives. And these modern mythologies allow us, once again, to construct a language about our origins, about ourselves, about others. But, as in the old cathedrals, it's never anything but our perspective, of course, our way of looking, in plain and simple words. And when they're too complicated, they seem to flash like SOS signals. An SOS on the display cases of museums. SOS, it's all too complicated! So we exaggerate, we stylize, we simplify, we explain. Finally … We explain what we've understood from it, and the story, the great story, the history of the world and of all humans, is told like a nursery rhyme. A nursery rhyme for adults of course, but a nursery rhyme nonetheless. Not that these stories are false, they're probably the most accurate that have ever been told: but the complexity of the world around us is such that it can be understood only in a stylized way. And we hear the story of the universe, the story of humans, the story of societies. And we hear it almost as a teaching, as a moral, like those for children. It's never real history of course. I mean, those nursery rhymes for adults are never The Truth. They're never a fixed truth. We researchers, we grope. We question. We deconstruct. We doubt. We perceive the immensity of what we don't know. Our history, the past, our past, is a recomposed past, fleeting, always, but sounding like a need to inscribe ourselves in time. In a very long time, until the first words of the first story, until the shores of the first humans and of the distant Big Bang, the beginning of all things. But that immense history is totally beyond us, those billions of years of thermonuclear explosions, of lives emerging from the most inert mineral, of bacteria that become tyrannosaurs, of dinosaurs swept away in smoke and of primates turning into humans. In its reality. In its concrete processes, this history is totally beyond us, all of us.

It's understandable only on condition that it's ultimately reduced to our understanding alone. An immense reduction, but a necessary reduction in the face of the ancient cult of relics, a cult so simple, so obvious, that it could be summed up in the recitation of a single book.

My thousands of flints from fifty thousand years ago are *my* book. My way of tackling this immense history through the small end of the telescope. Ultimately, after all, surely this history of the Great Whole can be modelled only in its beginning and its end, from the thermonuclear heat of the Big Bang to the universal Big Crunch? And everyone tries to fill in, at their level, what happens between these two explosions of infinite light. Everyone rubs up against the immense complexity of the realities of our little world. And for me, my small end of the telescope is a few hundred millennia. A very brief history, almost an anecdote, but not just any anecdote, because these little flints, in the drawers of the Peabody Museum, would become surprisingly familiar to me, like a copy-and-paste of something I knew perfectly well already. However, I had never directly analysed archaeological collections from Lebanon and the eastern flanks of the Mediterranean. These drawers contained thousands of flint points. But there are a thousand ways to make flint points, and these points aren't just any points. What I have before my eyes are precisely the technologies of the humans of the Neronian. The Neronian, that question mark that in 2004 I had placed over those astonishing small standardized flint points that I recognized in five closely grouped sites in the heart of the middle Rhône valley.

But these technologies are highly particular. What we have here are not resemblances but, clearly, societies that shared a sophisticated and quite singular technological knowledge. For more than eight months in the bunkers of this American institution, the elements of the equation are put in place; I analyse, I dissect, I decipher precisely more than sixteen thousand flints without finding the slightest major divergence between these technologies from the two shores of the Mediterranean. No matter what, it's the same thing. The same traditions. Precisely the same technological culture that I see emerging behind these productions of small flint points.

The problem is that in the Rhône Valley my Neronian is sandwiched in a vast archaeological sequence whose artisan is undoubtedly Neanderthal. But these points, here, on the slopes of Mount Lebanon, were made by *Homo sapiens*. More than three thousand kilometres separate the two sites. And these technologies are a replication of each other. You have to try and understand what I'm confronted with here. It seems complicated. It's always more complicated than we thought. But here, we touch on the question of the last Neanderthals and the colonization of Europe by *Sapiens*. No. In fact, it's quite simple.

We need to rewrite everything.

Nanotechnologies of the first humans

So *Sapiens* didn't arrive in Europe between the forty-fifth and forty-second millennia. So the arrival of *Sapiens* didn't mark the extinction of Neanderthal populations in an immediate, linear process. So *Sapiens* and Neanderthals actually shared certain territories in Europe. Not in Eastern Europe but in the west of Western Europe. So these ancient *Sapiens*, from the fifty-fourth millennium, were the bearers of cultural traditions so strong and homogeneous that we can follow them, recognize their very precise traces on trans-Mediterranean territories separated by several thousand kilometres. None of this is insignificant and these different conclusions open many doors to an understanding of the precise structure of these ancient societies.

These societies, the first to be confronted with Neanderthals on the European continent, had mastered remarkable technologies. Their flint crafts were surprisingly standardized, illustrating the existence of standardized productions, mass-produced, to within one or two millimetres. These artisanal designs reveal a certain way of being in the world. This way is ours. There is a simplicity in understanding the objectives of the artisans that we never encounter in Neanderthal flints. We can sense our own logics at play here. We can also sense very old forms of knowledge, very ancient traditions, rigid processes of the transmission of knowledge that deeply structure the organization of these populations. It would be impossible to discern the same cultural tradition three thousand kilometres apart on opposite shores of the Mediterranean if these populations were not structured, constrained, by powerful cultural servitudes. And the technologies for obtaining such flint points are indeed quite remarkable. Among the one thousand four hundred and ninety-nine Neronian points from the Mandrin cave, we can recognize two distinct categories of objects, combining large points, four to five centimetres long, and very small points. Technologically, these large flint points and these small points are absolutely identical, but the latter are in a Lilliputian mode: real microtechnologies from the fifty-fourth millennium. The reduction is sometimes so advanced that we can talk about Palaeolithic nanotechnologies. These flint nanopoints are less than ten millimetres long, at most, and all are at most two to three millimetres thick. A penny isn't very big. Remember a British halfpenny? It was really pretty small. Well, we could have placed two of our smallest Neronian nanopoints inside this halfpenny, without them sticking out over its edge. Amazing, isn't it? It's even far too small to make a serious toothpick. So what was the purpose of these astonishing nanotechnologies created by the first modern human beings in Europe?

Laure Metz, one of the students in the Mandrin team, would carry out five years of analysis in order to be able to work out the function of these tiny objects. Her doctoral thesis, defended in December 2015 at the University of Aix-Marseilles, led to conclusions that were unexpected, to say the least. These nanotechnologies are directly related to the weapon technologies developed by those societies. These tiny points are all weapons. Functional weapons, fractured and brought back to the cave on return from hunting, or from war. And these points weighing a few grams could have worked only when shot by a bow. Archery from the fifty-fourth millennium. A good forty millennia before the supposed invention of the bow across Eurasia. That's quite something. At that time, Neanderthals used heavy hand-held spears, also armed with flint armatures, but with armatures one hundred times heavier than the flying points of *Sapiens*. While the mastery of fire is often perceived as a crucial technological step, and even more as a step that fostered the processes of the socialization of our ancestors, archery doesn't represent a mere dialectical leap. Archery is another universe. It's advanced technology. A watchmaker's technology, involving an understanding, a mastery, and even more an association and synergy of materials with radically different properties, using the spring and the force of the wood fibre, constraining it with flexible and resistant fibres, associating it with composite elements, point, stock, ligatures, and fletching, whose weight and length are intimately correlated to the length and strength of the bow. Archery already involves an overall understanding of the properties of natural materials and their potential, if they are networked. Here we are confronted with degrees of projection of ideas, and levels of conceptualization, so relevant and so effective that these same archery technologies still, fifty-four thousand years later, occupy an important place in all societies across the world. While the French Archery Federation has nearly eighty thousand members, it is estimated that nearly twenty-four million people practice archery regularly in the United States. When they have not been swept away by the unbridled globalization of recent decades, these technologies still occupy a central place in the economy of many hunter-gatherer societies. These technologies, very much alive, distance us profoundly from the skills proven to have existed among the Neanderthals. And the right question is not to know if Neanderthals were also capable of doing this or that. They could well have walked on the moon if they hadn't died out – who knows? I don't. The right question is always to know what, very concretely, these societies achieved, or didn't. Once again, I don't know. I analyse. I question. I note. I scratch where there's something bothering me – where, generally, we don't like to look too closely. Here, I'm just telling it how it is. I note that archery technologies don't fit in with what we know about Neanderthal populations. At least until we discover a very old Neanderthal lunar base. While I rather like the idea in itself, until such a discovery is made, I'll just say that Neanderthals didn't actually go to the moon.

From impressionistic dots to thick lines

These *Sapiens* societies dazzle us with their quite remarkable technologies and we've deciphered their rigidity, rigour and repetitiveness across thousands of kilometres and beyond the seas.

These societies, powerfully rigid, do not seem to dazzle in their creativity, in their capacities for emancipation from their ancient principles. You shouldn't see anything paradoxical about this. These very advanced technologies are technologically impressive but they don't illustrate the emergence of any new technologies in the fifty-fourth millennium; rather, they probably replicate some already very ancient technological traditions. While archery doesn't seem to have any antecedents in Eurasia, it's possible to trace it beyond the eightieth millennium in a group of African societies. This was probably a new idea only on the Eurasian supercontinent. Our Neronians, their ancestors, merely developed a certain number of technological and cultural standards in association with these archery traditions – traditions that were already more than thirty millennia old at the time when we see these technologies emerging in western Eurasia. Obviously, I can't draw a direct line between these old African archeries and our Neronian technologies. These solutions could have been forgotten and reinvented on many occasions over time. Let's say that in the absence of patents, such ideas, such brilliant synergies, must have leaked and flared up like a flash in the pan across the continents, the moment the spectator of an innovation first laid eyes on the flight of the first arrow. But this is perhaps also an obviously simplistic, naive answer, and one that avoids confronting the complexity of the propagation of ideas. The question of the methods of diffusion of major technological innovations across the planet remains one of the worrying puzzles of our disciplines. No scientific article directly confronts this question of the emergence of technologies. The biggest international journals pick up the headlines; the oldest *this* and the oldest *that;* and the progressive ageing of our discoveries tiresomely allows for the increasing number of major announcements in the media. You see them regularly in online news: these first decorated caves, these oldest adornments, these first intelligent features. They're always older, every week, like when you drop a spool of thread and try to make it come back to you by pulling on the string: you just see the spool unwind endlessly without it ever getting closer to you. The first bow, the first art, the first arrow, the first fire, the first cut stone, is the spool of thread that unfolds endlessly without ever making the slightest movement towards us. We can well imagine that when we've pulled all the thread we'll finally see the end of the string and the spool on which it was wrapped. But, in the history of technology, all this is very uncertain. The first cut stones that for decades we thought were 2.6 million years old suddenly grew seven hundred millennia older in 2015, sending us back 3.3 million years, all at once predating the very emergence of the genus *Homo* and making us profoundly question the first technologies that we'd like to have synchronized with the simple and direct emergence of our biological ancestors. The question of the origin of all things is immensely complex, as much as the question of the diffusion of ideas, systems, concepts and techniques. Some technologies spread like wildfire and suddenly appear to us as universally known by all human societies, as an immediate common good that transgresses all global spaces in a single instant.

The short, simple, easy answer is technological leakage, exchange, analysis, copying. But, for us archaeologists, who question the origin and evolution of human things, the question probably also involves the inadequate resolution of our data and of our analysis procedures. All this is at the same time because we are so bad at assessing distant times. Defining which came first, the chicken or the egg, is always a challenge, especially when neither the egg nor the chicken are biological entities but ideas, concepts or technological knowledge. These ideas, however brilliant they may be, could well have emerged a thousand times and been buried a thousand other times. And, with tens of millennia of hindsight and without our being able to see precisely the origin of everything, it might well appear, from a distance, with the hindsight and the tendentious blur of millennia, like a formidable impressionist painting. A painting where each touch of colour seems to connect to the point next to it, thus adding an extra gleam to this rising sun – which is in reality just an addition of autonomous touches of colour, without continuity, without regularity. When we approach, finally, we see nothing other than these small lumps of colour, these small touches of paint. But, from a distance, everything makes sense. In reality this parable doesn't really work. In a painting, there's a painter, a project, a vision. In impressionism, there's an impressionist. A painter with the talent of an illusionist. But, in the origin of the bow and of all great inventions, there's no great project that transcends humans. There is merely the fact of their presence. Of their absence. We see these ways of knapping flint that are, in an instant, everywhere. But this instant, this synchrony, is an illusion. For a long time, we thought that certain ways of knapping flint were three hundred to three hundred and fifty millennia old. We call this 'Levallois knapping', as in the Paris metro station.[6] But three hundred to three hundred and fifty millennia still leaves an uncertainty of fifty millennia. Let's admit it. These fifty millennia are the thickness of the line of our archaeological resolutions, you see … But already, this beautiful resolution to within fifty millennia is just a beautiful illusion. An illusion that few archaeologists are fooled by, I hope.

A thousand millennia of Anatolian transgressions

Twenty years ago, I was leading research in central Anatolia, in the heart of Cappadocia, on the slopes of an immense volcano, the Göllü Dağ. At its summit, at more than two thousand metres, were the remains of a Hittite fortified city from the ninth century BCE. Impressive fortifications with columns supported by stone lions, contemporary with the twenty-third dynasty of Egypt. More than a million years ago the volcano spat out immense flows of obsidian, veritable cliffs of natural glass, ranging from the deepest black to translucent grey. These obsidians from the Turkish highlands would be used throughout the ages. The Hittites dug incredible vases out of this natural black glass. A thousand years earlier the Mesopotamians had collected the obsidian of Göllü Dağ to carve their magic amulets. In the tenth millennium the last prehistoric people used it to polish black mirrors. A few millennia later some hugely talented stone knappers, specialized artisans, extracted very large, remarkably sharp points that were probably as much objects of prestige as simple arrowheads. These objects, transforming natural glass, all speak of luxury, representation and the magical knowledge of these societies. By prospecting around these recent workshops, which, after all, are about ten millennia old, I found, lying there in a dried-up dell on the eroded slopes of the volcano, a horse's jaw. I climbed this slippery slope and guessed correctly that there were small fragments of black obsidian surrounding these bones. When I extracted them from the ground I found I had two small obsidian objects whose technologies were completely familiar to me. These two objects were Levallois blades, products of the famous technologies widespread in Neanderthal crafts that enabled one to configure the shape of the object to be extracted from a block of flint, obsidian or crystal. This could produce ovoid edges (Levallois flakes), or very slender edges (Levallois blades), or perfectly triangular edges (known as Levallois points). In this part of Eurasia, these technologies had disappeared about forty millennia ago, precisely when Neanderthals disappeared. This horse must have died more than forty millennia ago. I then opened a huge trench on the sides of the volcano, descending in seven seasons of archaeological excavation into this small piece of Anatolian desert, to a depth of almost eight metres. The fallout of volcanic dust, trapped in the thickness of this trench, would allow us to position ourselves in time. These ashes have very particular physiochemical signatures that allow us to attribute them to this or that volcanic explosion. These black pipings on the edges of my trench comprised superb sketches of the site of these explosive events. Anatolia is a land of volcanoes, immense volcanoes exploding in immense calderas. These archaeological seasons showed us that our horse, whose remains lay directly under a thick layer of black volcanic ash, was in reality more than one hundred and sixty millennia old. But, well beyond that, this vast trench recorded the passages through time of prehistoric populations. One hundred and sixty millennia at the top. But probably one thousand three hundred millennia at the bottom ... The traces of these artisans on the Anatolian highlands were unexpected. From this volcanic ash we took the first bifaces found in situ in archaeological levels in Turkey. And with them the whole procession of the oldest technologies. The technologies of ancient human beings. Long before *Sapiens*. Long before Neanderthals. What humanity? No idea, but there were bipeds wandering around in the steppes of Eurasia in those distant eras. And you can guarantee that we don't know them all. The whole procession of ancient technologies passed by at the very bottom of the trench, technologies whose names are echoed in the old industries of the African continent; bifaces, cleavers, choppers of various kinds, polyhedra, large flakes, those immense, fairly classic flakes of the early Palaeolithic. And there, at the very bottom, on the floors of ancient volcanic flows that cooled perhaps one thousand three hundred millennia ago, a superb, remarkable Levallois core. There, one, then two, then three large Levallois flakes. Caricatures of Levallois. Well, you just have to sit down. You have to think. You see, Levallois is the range of technologies invented three hundred millennia ago, give or take fifty millennia. But, well, my Levallois must be a good million years old. So, we'll sit down for five minutes, OK?

Right.

My Levallois is here, no doubt about it. No need to nitpick. There aren't any giant moles capable of mixing up our archaeological layers. Everything's in place. And the funny thing is that when an obsidian object slides into the ground, it immediately gets scratched, its surface unfailingly records its movement. My Levallois lithic core here is brand new. It looks as though it's just been carved. It was wrapped in this volcanic dust more than one thousand millennia ago and it didn't move again until someone dug eight metres, by hand, to bring it back to light more than a million years later.

Okay. We've sat down. We're all settled. We can start thinking, properly. Levallois technologies are not three hundred millennia old, or three hundred and fifty, it doesn't matter. They're simply three times older. That's all. Simply. We assess the situation. We question it. We accept it. Our visions, our concepts, are wrong. So much the better. We move forward. Superb Levallois technologies had been completely mastered probably more than a million years ago. It's so elegant, next to their big pebbles roughly dealt with in just three strokes, resembling the technologies of more than three million years ago. It spices things up. It allows us to think. We come across this kind of incongruous anecdote from time to time in the life of an archaeologist. We feed on it. We talk about it. We think about it. We don't necessarily publish it. We can't publish everything and what a struggle it is to get these things published in the major international journals. We go to a committee of experts, we split hairs, we blabber, we palaver, we chew the fat with other specialists. This process is fundamental and represents one of the best tools invented by our disciplines for displaying the progress of our knowledge. But reviewers are sometimes pinheads when it comes to their general level of culture and breadth of mind. So, obviously, we can't pass our time getting our discoveries into all the pinheads on the planet unless we stop research, real research, the kind that comes from the real extraction of data from the field. We publish everything we can. We make available the maximum amount of data possible to the international community, so that everyone can make progress with us, especially those who are not mere pinheads. But there's always a fraction left. A sometimes significant fraction that we're left with, that we try to pass on to our students, so that they can go further. So that at least they can know. This is probably what we call being an old hand. Having experience. It doesn't come from books or articles but from direct contact with our material. I have countless such anecdotes in my head. Countless unpublished tales, mine and those of my friends, who also enjoy digging up the earth. In these anecdotes we can see the most unexpected events. The most baroque. We see humans gathering meteoric iron to knap obsidian twelve millennia ago, or Neanderthals hiding the tool that was used to cut up the bodies of their dead in the crevices of the rocks. We see the oldest adornments in the world, adornments that we don't know how to describe. Those for which we need to sit down, for a while, to think even a bit longer before talking about them. We see many other things, just as entertaining.

Right.

The Levallois is more than a thousand millennia old, sometimes, but when you think about it, it's not all that surprising. In those ancient times, people were familiar with all the technological elements that showed the interest of those distant societies in organizing and predefining the shapes of their stone tools. Finally, this leap of seven hundred millennia is almost an anecdote in the history of technology because these technologies don't really mark a crucial dialectical leap. They're part of a certain way of conceiving the exploitation of rocks, a very generic way in fact – one that was most likely reinvented many times in the history of technology.

Let's return to the technologies of our first *Sapiens* in Europe. The very generic technologies that we see emerging here and there, seven hundred millennia apart, in no way characterize the technological structures of the first modern humans in Europe, structures whose rather remarkable rigidity inevitably suggests a cultural framework that probably weighed very heavily and seems to have shaped the organization of these societies.

Mythologies of the first modern humans

Let's get back to Mediterranean France and look at the first *Sapiens* of the European continent.

Their cutting-edge microtechnologies show the existence of very strong standards in these human groups. These are technological standards, of course, but they're not underpinned by any mechanical constraints. Neither the properties of rocks nor the activities in which these pointed implements are then used require the development of such rigid technological rules and constraints. This rigidity doesn't tell us about the materials or the nature of the activities of these populations, or about their knowledge, or even about their technologies. These rigidities tell us about constraints that these societies imposed on themselves. The constraints are to be found in the history of these societies, in their way of conceiving the world, in their desire to reproduce ancestral gestures. To reproduce them in a standardized way. These societies are clearly constrained by mental structures and powerful cultural rules. And everything happens in the background of all that: traditions, duties, taboos. You do *this* and you never do *that*, otherwise … Taboos are boundaries, limits and myths, of course. Myths above all. Among us humans, traditions have their reasons that are quite unknown to reason itself. Reasons that reason does not understand. Such is the weight of traditions. And these reasons are never rational, functional, or simply logical. They could well be so, ultimately; but when we use words, it's never: 'Don't go into the forest alone, or you'll get lost', even if the real reason, the logic of this prohibition, is indeed this. But when we put it in words it generally becomes this: 'Don't go into the forest alone, or the big bad wolf will gobble you up.' Behind the technological constraints, the equally rigid standardizations of technological traditions, ways of knapping flints, always in the same way, lurk those stories, those prohibitions, those nursery rhymes about the first human, about the first knapped flint – stories that relate how humans discovered the first flint point. To each his own Prometheus. To each his own Odin, suspended by one foot from the branches of the primordial tree to seize the knowledge of runes, writing, magic, and offer it to human beings. In these nursery rhymes it's explained how, in order not to offend the spirits, you have to make a certain gesture before extracting the flint point. It will then be an irresistible point, the one that always brings game back to the hunter. All human societies are made this way. And the more the craft traditions are standardized, the heavier the legacy – the more the legacy is rooted in history. Not really the history of humans, as we would define it, but rather in mythical history, in narrated history. The one that we repeat each time we make flint points. And that's what's important in these points from the fifty-fourth millennium. They open the door to conceptions of the world that are those of all humans today on earth.

They reveal that, already in those times, human beings were human. The important information, the great piece of information, is *that*.

All the rest is detail. Interesting, fundamental, remarkable, sometimes astonishing detail. But detail, nonetheless. We know that in the fifty-fourth millennium, while our *Sapiens* were wandering over the west of Europe on the very lands of the Neanderthals, at the eastern end of Eurasia entire populations were crossing immense stretches of sea to colonize Australia. The navigation tool – wooden objects, objects made of archaeologically invisible fibres – had been completely mastered. We know this because if there had been no navigation no population would have been able to cross these immense stretches of sea. We therefore only know it in negative terms: it's all quite invisible but we know it with certainty.

At the other end of Eurasia but precisely at the same time we know that our Neronians had completely mastered archery. Other woods, other fibres that have also disappeared. We know this through the very precise analysis of their arrowheads, rot-proof because made of flint, an almost eternal material. Again, we have certainty based on essentially invisible realities.

And we know that these humans structured their world, their imagination, their origins, like us, around stories and myths. We can discern it in the invisible, too, but we know it with the same certainty. I would have liked to tell you some of these distant stories, but I think they're well and truly lost. It's a shame, because they probably had a lot to tell us about the moment when *Sapiens* met Neanderthals in Europe.

Human beings known as shadows

The Mandrin cave in the Rhône Valley tells us about this first contact, pushing the arrival of *Sapiens* in Western Europe back by twelve millennia. But this step back in time is just another anecdote and doesn't tell us anything about the populations – about their origins, their traditions, their encounter with Neanderthals in these territories of the Western world. The event took place fifty-four thousand years ago. It could well have happened a hundred years ago, or a hundred millennia ago; the time would actually have little importance. The relevance is to be sought elsewhere: in the structure of those populations with regard to the Neanderthal aboriginal populations; in the organization of those societies; in the precise interactions they may have had with other societies on this same territory; in their geographical origin and even more in the structure of their technological traditions and what those traditions tell us about the fundamental structures of those human societies.

The question of time seems to have been finally answered, give or take a few seasons. These *Sapiens* are in Neanderthal territory and perhaps, for the first time in Europe, we can state that they could be in direct contact with the aboriginal populations, Neanderthals. We see here the only possibility of contact. This is already a notable advance but perhaps it was only a distant contact. A visual contact, where one figure, perhaps, looks at the other, secretly, without their knowledge, without ever revealing himself to them.

The analyses of our soot mean we can evaluate the precise temporality of this moment. Not to within a thousand years or so, but to within six months or so. It is probably less than a year that separates the last Neanderthal fire in the cave from the first *Sapiens* fire. One year, max. For the first time, the beginnings of an encounter. It is not yet taking shape, there's nothing in the reduction of time, from a thousand years to a few months, that would allow us to glimpse a contact between these humanities. We have simply gained, on two millennia of uncertainty, one thousand nine hundred and ninety-nine years of chronological precision. That's already not bad and, methodologically, it's a real feat. Uncertain time seems to finally collapse on itself. From now on, only a few months could separate our two humanities. The meeting is within reach but it's still not taking shape. Let's say that, at this stage, and for the first time, there's a hint of it. The improbable meeting of two humanities separated by half a million years of parallel evolution. We knew that this meeting had existed, but we only knew it in a negative implicit way, like something unsaid, that can't quite be seen, that we don't know how to describe. We don't know how to imagine it either, so inexpressible is the meeting and so far from all our mental schemas. We knew of it only through the analysis of our genes, which still carry in our flesh the trace of these contacts. We also thought we knew of it through the traces of cultural transmissions, which can apparently be discerned in certain crafts of the last Neanderthal populations.

But did these humanities cross paths somewhere on the European continent? Or should we consider processes of avoidance that were powerfully anchored within the Neanderthal populations and that drove them to flee all contact with the newcomers? That drove them to make themselves invisible in their own territories? Yes, I know, it may raise a smile, but wouldn't this smile be one of the many indicators of our inability to comprehend this astonishing moment in the history of humanity? Can we envisage the existence of entire populations deciding to make themselves invisible to a different humanity settling in their territory?

The first contact would not even have been visual: it would have been a contact made in their minds, in the certainty of being on an occupied territory but never seeing the occupiers, recognizing only their traces, sometimes, in the caves where they settled. Recognizing the remains of this fire that had still been burning not long ago. Seeing nothing but the shadow of others.

Human beings thought of as *shadows* of others.

Such a scenario is so far from our own behaviour, from our own history, where the encounter of our society with different civilizations has always been direct, frontal and brutal. Examples: the Gallic War, the Crusades, the confrontation with the Mongol Empire, the coloniz-ation of Australia, Africa, the conquest of the Americas. In our history we are used to stomping in with our big boots. And we end up making our boots echo far beyond our solar system. Towards the unfathomable galactic horizons. In the golden discs, these engraved messages placed on our first interstellar probes in 1977. The messages of Voyager I and II were addressed to the void of infinity eight years after we planted our standards on the Moon on 21 July 1969. It is difficult not to notice us in our travels, as it seems to us perfectly natural to claim a presence everywhere and have each of our steps acknowledged.

How could we then imagine human beings as *shadows*?

All this would be so far removed from the cultural structures of our societies that we find it difficult to consider seriously the existence of profound processes of avoidance. And yet, the question is actually far from absurd. During the conquest of the Americas, the colonists were commonly avoided, observed without ever revealing themselves. Often the explorers felt watched, knew they were in occupied territories: but avoidance was commonly the rule – the first strategy of the aboriginal populations faced with these strange travellers. And don't imagine that this shadowy tactic was just a simple avoidance strategy, a temporary ploy, to give themselves a little time, a few days, a few weeks, a few months, to understand the strange strangers before meeting them. More than four hundred years after Christopher Columbus and Hernán Cortés, on 29 August 1911, J. B. Webber, the sheriff of Oroville in California, got an early morning call from the corral of a slaughter house saying that the butchers there 'were holding a wild man and would he please come and take him off their hands'.[7] The unfortunate 'wild man' had come to die among the white devils. He was the last of his people, the last Yahi Indian, from a population whose very existence was unknown. Their language was unknown, too. Theodora Kroeber powerfully recounts in her book *Ishi in Two Worlds* the epic of this man whose tribe had voluntarily remained invisible for centuries. Ishi would see his people disappear one after the other. One fine morning Ishi was the last of the Yahi. His mother was dead. They had all died in their retreats. Died in their invisibility, in the eyes of those who now in their millions occupied their former territory. He was alone. He would remain like this for three long years, hidden from 1908 to 1911. But it had to end. And, on this August morning in 1911, Ishi walked toward Oroville, the strange city of the Whites, to meet his own death. One does not live alone forever. One does not live for oneself. If his last years of life were among the Whites, he never fully understood them, any more than the Whites could ever truly understand Ishi. His last communications were still lovingly imbued with the modesty and prohibitions of his own traditions – traditions of which we knew nothing. His very name has remained unknown

26

to us. Ishi, the word he gave to name him, was not his name. Ishi meant 'human'. Contact with otherness, with difference, is always such a step into the unknown, whether it was one hundred and eleven years ago, or fifty-four millennia ago. But perhaps these one hundred and eleven years still seem a long time ago, too long ago to be perceived as a real story and not as those beautiful fictions that we tell ourselves for entertainment's sake, like the magical and bizarre stories that lull children to sleep. Ultimately, 1911 was already a Wild West. We can hear the horses and the carriages and we can make out those cowboys under their beautiful Stetsons. We are anchored in our imaginations, in our mythologies. 1911 is already a long time ago, a distant, extinct period, where history has already been replaced by myth. The Indians and cowboys, the conquest of the West, the last gold-rushes. The long time before the great industrial wars in which flesh was crushed rigorously, row after row, in orderly batteries. The time before. So, forget those nursery rhymes of times gone by. In reality, the Indians still exist, and we know that there are still some, invisible, in the last great wild forests. We see videos, from time to time, here and there on the World Wide Web, showing us a helicopter being attacked by naked, painted warriors, riddling the flying ship with arrows. These videos reveal to us the last of the last wild humans who have never had contact with 'civilization'. But, if you look closely, there, furtively, on the edge of the video, you will see a basin that is too grey, too shiny, probably made of aluminium. The dented metal doesn't reveal the transformation of bauxite by these Amazonian populations, but it does show us that these 'wild' people are not unaware of the forest eaters. They obviously know them too well to approach them. These men, too, of course, have become *shadows*. They're not firing arrows at the helicopter because they fear that astonishing iron bird but because they know all too well that they no longer want any contact with the Whites. One must read and reread Lucien Bodard's *Green Hell*[8] to get even a slight inkling of the industrial massacres perpetrated not in the time of children's nursery rhymes, or in the time of Cortés, or among the last cowboys, but in the 1960s and still in 2022, under Bolsonaro. '*O índio mudou, está evoluindo, está cada vez mais um ser humano igual a nós*' ('The Indian has changed, he is evolving, he is more and more a human being like us'). His generosity will be his downfall … The hideous sentence dates from 2020. It smells of the putrefaction of a re-education camp and

27

can be contextualized in remarkable ways by Bodard's writings. What is still happening in the great Amazonian forest, before our impassive eyes? The book is unspeakable. It's the cutting of flesh. Cutting in the literal sense. Permanently. It's slavery before the cutting. It's all the most unimaginable and disgusting horror that Bodard witnesses. And, if the men often simply lie down to die in despair, also, as a last hope, they often flee, and become *shadows*. Sometimes they flee before any contact, when a mere rumour announces the coming of the Whites, or of the half-breeds who are often worse than the Whites in their terrible desires, as uprooted humans, to be recognized by their half-fathers as real Whites. Sometimes they give up and understand that only invisibility still represents a small hope of life.

> The hostilities had a complex plot. It was the Memcromotires who attacked. They are experienced killers, but semicivilized. They have had a relationship with the whites for a long time. The latter have often used them as their eyes in the jungle, as guides and spies, and also to 'contact' wild tribes. [...]
>
> The war among Indians is also a way of getting rid of Indians. Moreover, the white adventurers of the Arinos were particularly anxious not to be further annoyed by the Kraiacaras, who are a very individual people, very distrustful, and very much in the way. They were a special branch of the Capajos; descended from half-castes who had been absorbed by the jungle, forgotten, degenerated, and Indianized; they had become Indians opposing the new half-castes of the present time.

But the Mencromotires were too numerous, and their setbacks served only to increase their rage. They were determined to kill all the Kraiacaras, or to make them move elsewhere, far away. As it happened, the Kraiacaras ended up by leaving. But where had they gone to? Had they escaped to a distant jungle hide-out, or were they going to reappear suddenly and dangerously where they weren't expected? A certain fear hung over the upper Xingu and, as always in Amazonia when one kind of Indian is dreaded, the rumour spread that they – in this case the Kraiacaras – were gigantic specimens, over six feet tall.[9] And we realize from Bodard's words that it has become impossible for us to understand these realities of our world because they are so far away from us. They sound so much like those children's stories, those nursery rhymes describing how the elves disappeared from Brittany or Ireland. The little people of the moors transformed, becoming so small that no one can see them any more, and then disappeared from our eyes, completely – we don't really know if they still exist. Or if they ever existed, for that matter. Doubt hangs over us, but we fear them. We dread them. In Ireland, Iceland and Norway, at certain times people still leave them something to eat and drink, something to welcome them in by the hearth, to solicit the good luck they can bring.

So, could humans disappear like the elves? Become *shadows*? Physically transforming themselves into invisible giants or elves of the mists? The story is fascinating. And it's probably true. This is probably how humans die, by slipping from history into imagination. From their reality into our stories, into our myths.

Let's get back to the most basic data. The singular moment of contact occurred at the crossroads of two major disciplines on which archaeology draws: physical anthropology and cultural anthropology. To know if we are getting back to the first contact, we need the biological trace of these humanities. Their bone remains, or their genetic signatures. And we need to determine the structures of their technological traditions – the precise technological relationships that can be drawn between these crafts and the other contemporary traditions recognized across Eurasia. We must try to determine how far these traditions were affiliated with the remains of other human communities in the space and time surrounding them. It's from this structural approach that meaning can be extracted, never in the anecdote but in the intertwining of the events that made it possible. In the intertwining that shapes its distant historical processes.

A first contact with inverted values

It seems possible to characterize not *the* first contact, but *a* first contact, in Mediterranean France. It's easy to understand that each cultural group must have been confronted with very particular historical processes. The Neanderthal societies of this small Mediterranean area were probably not confronted with the same sequence of events as the distant Neanderthals of the Siberian Altai. Even very locally, the consensus now seems to be that the neighbouring populations of Burgundy, to the North, or of the Po Valley beyond the Alps already show us the existence of very distinct societies and historical trajectories. They are certainly Neanderthals – the last Neanderthals but their traditions differ profoundly. And their meeting, their first contact, if this – the first of the last contacts – did indeed take place, must tell us other stories. Other processes. My research in the Rhône Valley therefore provides us with access to only a small story. A modest story, one that perhaps doesn't really tell us much about history. But a story which does perhaps tell us about one of the most important meetings in the history of these two humanities. An absolutely critical historical moment in which our humanity will spread over an unknown continent and another humanity will definitively die out. A very small window through time.

The window that it opens onto in this region is all the more exciting because it is in one of the most important migratory corridors on the European continent. Europe is geographically highly segmented and the Rhône corridor represents one of the main natural axes allowing the Mediterranean areas to be directly connected to the steppe expanses of northern Europe. It's in the heart of this very particular migratory space that we can discern the oldest arrival of *Sapiens* populations on the European continent, so this very particular moment happened right back in the fifty-fourth millennium, a good ten millennia before what was envisaged before we published our results in February 2022. Make no mistake, those ten millennia do not resemble ten millennia of exchanges between *Sapiens* and Neanderthals in Europe but a simple pulsation of about forty years, almost an anecdote, during which populations of modern humans settled for a time on the opposite side of the Mediterranean. We have here a first, singular contact. We can identify this time on at least three distinct scales. The time during which this event takes place refers solely to the hypothetical time of physiochemical measurements. This fifty-fourth millennium is a chimera that we set up so that the mind can connect to a precise point, like a lighthouse guiding lost ships from afar. This fifty-fourth millennium actually extends, statistically speaking, from fifty-two thousand to fifty-seven thousand years ago. Over five millennia. These five millennia are precisely the thickness of time that separates the first footstep on the Moon from the first Egyptian dynasty. It is therefore not a time on a human scale, nor even on the scale of cultures and societies as we understand them – even if, within populations with an oral tradition, human memories, transmissions and traditions can perpetuate themselves well beyond five millennia. But written societies seem more fragile in terms of memory, more subject to forms of amnesia than oral populations.

Finally, we must probably yet again reverse our views. The spoken word remains while the written word flies away.

This physiochemical time is of only very secondary interest, then, and doesn't enable us to touch any of the realities of these human societies. But, in these archaeological levels of the Mandrin cave, two other times have been identified – drawing on the analysis of our soot. These two other times are times on the human scale. Identified by traces of their fires, they are literally human times.

The first time is that of contact. That of contact between Neanderthals and *Sapiens*.

The second time is the duration. The precise duration of the *Sapiens* occupation in Neanderthal territory.

These two times are absolutely crucial elements in these equations with several unknowns. I won't dwell on the analytical methods developed by my teams using the techniques set out by Ségolène Vandevelde. Let's simply keep in mind that if these methods are remarkably complex in their implementation, they are simple in their principles, and this gives them their robustness and unexpected precision. Here, without further ado, is what these human times are. Six months. And forty years.

Plus or minus six months, this is the time that separates the last Neanderthal fire in the cave from the first *Sapiens* fire. It's a maximum time. It here includes the possibility of a 'zero hour' when *Sapiens* and the creature would have crossed paths in the cavern. Or not. We're sticking to six months – let's keep to this one certain figure: after all, if we make an appointment for six months, we could well meet, but the possibility of having dinner together, just the two of us, still seems rather improbable. Well ... It *could* work, who knows? Let's make an appointment!

Forty years is the period over which these *Sapiens* settled in Neanderthal territory. For if these two humanities may never have crossed paths, it's a certain fact that these first *Sapiens* of the European continent are indeed in Neanderthal lands. And this territory will remain Neanderthal for a good dozen more millennia. The analysis of the soot left by their fires on the walls of the cave shows that they will return to the cave each year during these forty years. Forty years is a significant period and the recurring occupation of this site underlines the fact that this is far from a trivial matter. We're not looking at a small encampment of a few days or a few weeks. We're not looking at a handful of hunters advancing far into unknown lands that they've come to reconnoitre. This duration corresponds to a human life and, obviously, for what is already quite a large group. A small handful of human beings can't settle for forty years in a distant territory. Our studies show, moreover, that the tooth of the little *Sapien* found on this archaeological soil is that of a young child who must have been between three and six years old. The associated archaeological material discovered is extremely rich: it's not just a few scattered tools abandoned by a handful of horse and bison hunters. It's thousands of objects, an incredible jumble of all their flint crafts. The archaeological implements abandoned by this population is quite remarkably varied. We're looking at a group of women and children and what is probably already quite a large number of individuals, large enough to be able to organize this population over a large territory around the cave – for several decades. This territory can be clearly seen taking shape through the circulation of flints. The flints have precise geological origins, sometimes geographically very localized, commonly over a few kilometres and sometimes over only a few hundred metres. This is precisely what is astonishing here. This group of *Sapiens* from the fifty-fourth millennium who settled for about forty years in this vast territory of the middle Rhône valley knows this territory very well. I mean, they know it *too* well. Too well for a population that has been settling in unknown lands for only forty years. Many of these flints come from a territory lying within a vast radius of thirty to one hundred kilometres around the cave in all directions, from Gard to Ardèche, from the Vaucluse mountains to Vercors. All the resources are known and exploited, even the rare rocks from deep, isolated valleys, whose resources are visible and accessible only over a hundred metres. How is this possible? Neanderthals knew these resources perfectly well; these populations had occupied these spaces for all eternity, for tens of millennia. This very precise knowledge of the land had been passed down from Neanderthal mouths to Neanderthal ears from generation to generation. The vast territory outlined here corresponds to several thousand square kilometres and it's difficult to see how a population could have combed the entirety of such a land in a single generation. Getting to know the tiniest resources in record time.

It all gets a bit more puzzling

When we analyse the Neanderthal territories visible in the Mandrin cave through time, from the hundred and twentieth millennium to their extinction forty-two millennia ago, we can make out some very distinct craft traditions. The Neanderthals of fifty-five thousand years ago did not have the same technological traditions at all as those of the fifty-second millennium, who themselves did not have the same know-how as those of the fifty-first millennium. We are faced with societies imbued with very distinct and quite easily recognizable styles, cultures and forms of knowledge. And, if the crafts are clearly recognizable, the territories and the choice of the rocks that were mainly worked also evolve considerably. In each cultural phase, territories and very particular ways of exploiting their resources emerge. The problem is that here a very clear continuous transmission of knowledge emerges between the last Neanderthals and the first *Sapiens* of the fifty-fourth millennium – within populations that are precisely separated by six months. Not only do the *Sapiens* know every resource over thousands of square kilometres but, even more astonishingly, the categories of rocks favoured are precisely those that were favoured within the Neanderthal group preceding *Sapiens* in this cave. And this particular relationship to rocks and territory is not found anywhere else in the ten other levels of the archaeological sequence that spans eighty millennia. This relationship to rocks from a territory of several thousand square kilometres clearly unites these last Neanderthals and these first *Sapiens*.

Here we touch on a very particular process of transmission. A transmission from Neanderthal to *Sapiens*. And, if we have pondered the thousand ways in which Neanderthal societies seem to have been profoundly transformed upon contact with the first European *Sapiens*, the reverse process has never been clearly identified. Never considered or never documented.

Making the best of a bad job

When we discuss this encounter, processes of cultural transmission between the two populations are being systematically taken into consideration. But they are always unilateral, with *Sapiens* bringing new technological light to Neanderthals who see their old technological traditions being swiftly transformed. But how can assimilations, transmissions, acculturations be imagined without reciprocal acculturation? As if these processes could be expressed from afar, with the arrival of the first *Sapiens* in Europe being preceded by a rumour, a bell ringing to mark the end of the Neanderthals' playtime. A powerful breath sweeping the old world and preparing the advent of modernity – a breath so powerful that it preceded the very passage of the *Sapiens* in their Westward march. A clash of civilizations – but without a clash, since it impacted on just one civilization. Acculturation before extinction, as if Neanderthals had nothing to offer to newcomers. And here, we are faced with the astonishing paradox of a cultural metamorphosis without genetic contribution in the Neanderthals and of genetic mixing without cultural enrichment in the first *Sapiens*, since this is what genetics reveals to us on the one hand, and the analysis of the cultural traditions of those distant societies on the other. But are we really sure of our connections, our constructions, our conclusions?

In reality, the human remains of this pivotal period of our history are so rare that the actors of this strange opera are no longer really discernible. As if the actors were no longer part of the stage or the setting. We would have to find, in each place, both the arts and the artisans. The tools, the weapons and the bodies. But, while we have millions of their abandoned objects, we generally do not have the associated bodies. Is it really Neanderthals who leave these objects here? These old human remains are incredibly rare. You need a lot of luck to discover a few teeth or a bit of more or less altered DNA here. And, throughout Europe, for the three almost complete *Sapiens* remains the context isn't there. In one place, we have two human skulls wonderfully preserved in the depths of a karst region in Romania. In another, a woman's skull remarkably preserved in the heart of a cave in Bohemia. They are among the oldest remains of *Sapiens* populations on the continent. And we have almost nothing to make them speak. Were they carried here by the flows of underground rivers? It's impossible to associate them definitely with a very specific archaeological context, even just a single object abandoned by their society. What skills did these populations have, what traditions, what was their group, their technologies, their history? We have almost nothing to connect these three remarkable remains to the other archaeological traces of the first *Sapiens* in Europe. Without a precise context, they can no longer speak to us. The organization of their societies can no longer be reconstructed. These remains are clearly more than thirty millennia old. But their age is uncertain and the measurements established are probably minimal. These bones could well be forty thousand years old. But couldn't the skull of this Bohemian woman be more than fifty millennia old? The conclusions are in reality impossible. The skull discovered in the 1950s was solidified using hide glue. The bones are so impregnated with these organic compounds that when the teams try to restore the genome of this lady, they manage to restore the complete mitochondrial genome … of a cow. The unfortunate beast whose remains had been used to make this hide glue seventy years ago was now genetically immortalized by the palaeogenetic teams of the Max Planck Institute in Leipzig.

So, we compensate a little and we make the best of a bad job by carrying out more and more physiochemical analyses. We try to fill the gaps using genetic, radiometric and isotopic samples and measurements. All the knowledge of physicists is put to good use. We know nothing of the traditions of these ancient people or of their journey but genetics outlines an encounter, a few generations earlier, with Neanderthals. The event took place a hundred years, or two hundred years, or three hundred years, before the birth of two of these individuals. Something like that, probably. But these traces tell us nothing specific about this little story. About the raw reality of these encounters. About these loves, these encounters, these rapes, and all the other stories written in the genome of these old *Sapiens*. And is there anything else apart from these skulls?

A hundred and fifty years of archaeology have led to the discovery of seven teeth and a little DNA. For all of Europe. It doesn't add up.

Neanderthals don't do much better, they're just as discreet. Although we know of about twenty Neanderthal remains spread between Belgium and Croatia and directly dated to between the fortieth and fiftieth millennia, all come from ancient excavations and were found between the 1970s and the middle of the nineteenth century. Only six of these deposits have been able to yield directly dated Neanderthal remains using the best collagen purification methods developed in recent years by the

Oxford laboratory, probably one of the best teams in the world when it comes to carbon-14 dating. And the majority of these bones found in the past do not have a robustly defined archaeological context.

This sets the scene for the set of actors in one of the largest replacements of humanity ever recorded. And, across continental Europe, archaeological research is still struggling to find the slightest significant vestige of these distant populations. The Archaeological Grail here would not only be to find a skull or a more or less complete body – a mere tooth or a fraction of DNA can sometimes make the difference – but to be able to draw some robust connections linking these bones and these populations to their crafts, to their traditions. To the concrete material traces that are the only ones that define the real historical processes at work during this pivotal period of change and extinction of humanity. So, for lack of context, the hard sciences take over, try to extract the slightest information from these rare bones. To make the bone speak for want of being able to make its dead owner speak. To make the dead person speak under constraint, as one would make a prisoner speak under torture, or even the cow that glued those old skulls together ... But can all the knowledge of physics and chemistry combined really compensate for our inability to define human societies and historical processes? Probably not, and here more than ever we must make the best of a bad job. But what would happen if we could cross-reference the social and cultural structures of precisely defined societies with the discovery of more or less complete human remains? If we could work not on human bones discovered in 1970, or in the nineteenth century, but on such remains at the very moment of their discovery?

Suddenly confronted with the discovery of this archaeological grail in August 2015, I would have to face up to the limits of the method. Of almost all of our methods. Hoping finally to acquire a little certainty without too much effort, I would simply be plunged even deeper into doubt. Doubt in everything. As if nothing worked. As if this last Neanderthal refused to be put in a box without putting up a fight. And, from these mists, a few possibilities will emerge. But they are possibilities that raise questions – that open out onto other doubts. All unexpected. All astonishing.

Thoughts ... uncertain turbulences that remind me of the way these humans died. Or of the way humans in general die? We need to ponder the matter a little more.

36

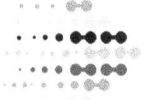

2

The last Neanderthal?
An improbable encounter

Oh no! It can't be true! But what the hell was he doing there? On the ground between this boxwood and this old oak tree?

I'm flabbergasted.

For goodness sake! You don't find a Neanderthal body by taking a stroll through the forest, just like that, lying on the side of the path. It's crazy. It's been over forty years since an even remotely complete Neanderthal body was found in France. And you don't discover the remains of the creature so easily. Well ... That's what we thought, until now. We're not even in the cave but outside on the slope, almost four metres beyond the opening of the small cavern, in the open air, like that, under the stars for ... what, forty-two millennia? Less than fifty millennia in any case, that's obvious. But I still don't know anything specific about the remarkable remains that have just appeared after a simple brushstroke on the edge of the path leading to the cave. A hell of a brushstroke, which has merely dusted the ground and removed the dried leaves from the scrubland. 'Ludovic, come and take a look.' On the site, someone's just called out to me. I'm right next to it, under a blazing sun dusty with the ash from the caves. And I can see it clearly. I can very clearly see five teeth. 'Oh wow. Uh ... Bravo. Okay, fine. Let's stop everything.' It's time to take your time. To sit down. To reflect. To think. A bit like that time when I was faced with the volcanic glasses involving technologies that were supposed to have been invented only seven hundred millennia later. When you see this kind of thing, and it happens from time to time, every ten or twenty years when you do a lot of fieldwork, you have to start by taking a step back. As usual, when faced with such an incredibly improbable situation, my first action is to freeze the moment. To immediately freeze the moment. You have to stop time. Freeze space. Completely control the action. The archaeological dig is stopped dead in its tracks. There's is no point in rushing, our little man has been waiting for a good handful of tens of millennia. Yes, our little man, because you don't lose five teeth so easily all at once. And, above all, they're within a few centimetres of their original position, molars and premolars aligned as they were in their owner's mouth. You can lose an arm and continue on your way. You can lose a leg. Some populations also experience voluntary mutilations, including every kind of amputation, sometimes ritual, sometimes marking infamy and punishment. Finding a foot or two, an arm or two, is no guarantee of the presence of a body. Amputations of bodies have existed at all times and in all societies. They still do. In Japan, among the Yakuza – a powerful mafia group claiming to be descended from the ancient Samurai – amputation of the finger is a way of permanently marking the man who has broken the rules. In New Guinea, among the Dugum Dani, little girls have to amputate one or two fingers to appease the spirit of the parent who has just died. In ancient China, during the Shang and Zhou dynasties, over the period of approximately one thousand five hundred years preceding our era, the punitive practice of *yue* consisted of cutting off legs or feet. In *yue* the cut could go right up to the middle part of the femur. No fewer than five hundred crimes could entail *yue*. The left leg was cut off first then the right in the event of a repeat offence. Such practices have sometimes been mentioned to account for the mutilated hands that are very commonly found in cave art, without it being possible to demonstrate the existence of these ritualized forms in prehistoric societies. But you can't amputate a large portion of your mandible without leaving your skin there and these five teeth most certainly mark the death of the individual. However, the dead person's body may have subsequently experienced many grim vicissitudes. The Gauls of southern France decorated the porticos of their cities with nailed skulls welcoming visitors. Perhaps this was a mark of infamy for the defeated enemies and a sign of power for the victors. Admittedly, this probably has a deeper impact than the sight of the red and white sign asking you to drive slowly when entering an urbanized space. And you did indeed have to slow down in order to analyse this striking signal. With a body, bits of bodies, the remains of our flesh,

just about anything is possible. But clearly the discovery is of signifi-
cance. And yet there's something that complicates the situation. The
remains lie on the ground, in very crumbly sand. And to understand
the meaning of these bones, we must find the precise position of each
remaining fragment. Preserving the original location is fundamental. Is
this tooth flat or vertical, with the enamel on top or bottom? The very
precise position of these fossils in the ground, to the nearest millimetre,
to the nearest degree, has something to tell us – but good heavens, it's all
so crumbly that if I even lightly brush over it everything might well get
shifted. I won't be able to restore the absolute position of each element.
And we've waited nearly forty years to find such a body in France. The
slightest bit of information about it is infinitely precious and another
encounter with such a creature is unlikely.

Well … We have to think a bit more … And once again, you don't
think with your face pressed up against the Neanderthal tooth. I remain
seated, looking at the immense landscape that opens up before me.
The huge Rhône Valley, wandering between the last elevations of the
Massif Central and the Alps. The bison, the horses, the mammoths and
the rhinos have been replaced by endless lines of mechanical machines.
They weave their way without end through the defile below the cave.
Everything goes through it. National and departmental roads, the
motorway, the TGV: these days seventy per cent of European traffic on
the North–South axis gets stuck in this bottleneck. People head towards
the beaches of the south just as in the past they followed the immense
seasonal migrations of large herbivores. People seek sand and sun just as
in the past they sought the great grassy meadows. Up and down it goes
in ceaseless motion. Ever since the Rhône has been the Rhône. And this
vast migratory corridor is the second biggest in the Mediterranean after
the Nile, which has never really been explored as fully as its importance
merits. We are far from the so-called classic region of prehistory. Far
from the Dordogne, from the Périgord, from the 'Land of Prehistoric
Humans' as it is sometimes called. And it was facing this same landscape,
with the immensity of the Rhône all before him, that in 1976 François
Bordes, the great twentieth-century prehistorian, suddenly exclaimed:
'The Dordogne! It's cat's piss compared to the Rhône!' This is a pictur-
esque way of putting it, and one particularly evocative of the respective
place in the ecosystems of these two geographical entities. And, indeed,

this immense migratory furrow is on the same scale as the great Mediterranean, that improbable geographical space, the only intercontinental sea that all by itself unites the entire ancient world, from Africa to Asia, coming to die on this small piece of European land, the Western dead end of the vastness of Eurasia. Last stop before the salty expanses of the Atlantic. And it's true that when looking at a map we get a good view of how the lovely lands of prehistoric humans appear as this dead end at the end of the world, one of the last corners of the West where life is good, but definitely not where peoples pass, cross paths, trade, rub shoulders roughly as they bump into one another, creating a more original human experience each time. Here, in this immense furrow, one of the main watering holes of the Mediterranean, this sea 'in the middle of the land', the *Mare internum*, everything is at stake. Everything has always been at stake. And not many people have come to look. To look in order to understand. To try to tell this story of where people pass, heading north. And heading south.

And these trucks that never stop passing by.

Well … We thought for a good while. I can raise my eyes from the immense landscape and again look these astonishing remains almost eye to eye.

I know what to do with the remains of our little man. It's simple. We're not going to use a brush. Too risky. We're not going to scrape. Too dangerous. 'Ah? But how're we going to do it, then?' exclaim the team members. We're going to remove each grain of sand. Grain by grain. One after the other. With tweezers. The team is stunned. But it's the solution. The only solution, I'm convinced. Let's go. Let's get on with it.

The operation will last seven years. Two months per year. It will reveal thirty-one teeth, the mandible, fragments of the palate and skull and series of phalanges from the left hand. The bones have hardly moved since the individual's death. The bones might have been carried down the slope in front of the cave, spread out for tens of millennia over hundreds of metres, in a slow mineral crawl, as if the creature had been crawling for eternity after having stopped walking. But no, it lies there, frozen in time. As if these forty-two or fifty millennia hadn't existed. As if time had stopped. Even here, outside the cave. He lies impassive, surrounded by rich archaeological furnishings where we find mammoth and bison and a jumble of sometimes wonderfully manufactured flint objects. He sleeps where his group abandoned him. If, that is, he wasn't the last of his group. Yes, because where I find him, at the top of our archaeological records, in very recent levels for our dear Neanderthals, he could well be the last Neanderthal. No more, no less. We decide to name him Thorin in reference to the works of J. R. R. Tolkien. Thorin is one of the last dwarf kings under the mountain … Neanderthal, as an amazing creature, already completely belongs to pop culture. And it's in this imaginary guise that I wish to pay tribute to the population that became extinct a few dozen millennia ago. Here is our Neanderthal, among the last dwarf kings.

But we're going to have to do a bit more work to understand Thorin. The millions of tweezer strokes were a minimum condition. Not a sufficient condition to understand the story of this man, the first such found in France since 1979. He seems to have been wedged alongside an old stone wall, positioned in a small basin, small but deep enough to contain a body. The ground wasn't dug, it's a natural basin and it was apparently there well before this body was deposited. Difficult to interpret. Is it all a matter of chance? Or was the body carefully deposited here by his people? And then there's the large stone deposited alongside his skull and which geologically has no place here. And the superb flint blade, the most beautiful among the thousands of stone objects found on the one hundred square metres excavated around this body; the flint lies a few centimetres away from his hand. Is all this the result of chance? A construction? *My* construction? Am I seeing things, as in a Rorschach test? Or am I really confronted here with a Neanderthal burial? All this is factual, a matter of observation and at this point you can reconstruct a moving ceremony of this people protecting the body of their dead kinsman in this hollow, between rock and basin, covering it with these few large blocks, abandoning it along with the most beautiful flint objects. You can also suggest that its location is random and that the only ceremony that ever took place is the one you are constructing by weaving uncertain connections between each of these facts. Are we accessing the creature's imagination and rituals, or are we limited by our projections, our fantasies, our interpretations, always too obvious to be true? Are we here confronted with the acts, the history, the consciousness of creatures who died more than forty millennia ago or are we constructing a reality that fails to mirror what really took place at the entrance of this cave?

And does this reality exist? I mean, is there an objective historical reality that is free from the observer? Or, as in quantum physics, does the observer here also profoundly influence the outcome of the experiment? As I write these few lines, I have no firm conclusion to offer you. I have no idea. I'm still unsure. In fact, the body was covered with thousands of flint objects, often waste products, as if the place where the remains of this individual were deposited had then become a dumping ground for rubbish. Did they simply treat the bodies of their dead as rubbish to be jettisoned? Living objects, dead objects. The dead with the dead and millennia pass and oblivion engulfs them. Wasn't this superb flint blade simply one of these thousands of abandoned objects? Only a clever person can answer that one. For now, we still have to dig, explore and get the tweezers ready for another ten years or so ... We'll see.

But I now face further fundamental questions. How old is this Neanderthal? It's a tricky one. He's lying in a basin and two hypotheses present themselves to us. Either the basin was already there when the individual died. Or this slight folding of the ground took place later. Depending on the interpretation, our individual could be between forty-two and forty-five millennia old or between forty-six and fifty. That's already a good margin. We are still oscillating between two geological layers, B or C, without any real certainty. But don't panic, no worries: all in all, to within four or five millennia, our individual would still remain among the best-dated Neanderthals in the world. The multi-millennial range may provide the remains but for we Palaeolithic specialists the chronological constraint is already remarkable. Lucky is anyone who finds a Neanderthal. Lucky is anyone who can give that Neanderthal a robust age to within five millennia ... You see, when I tell you that doubt is our everyday friend, well, we get used to it ... In any case, the cultural framework of the population to which he belonged is well defined. Layers B and C belong to the same cultural group, known as post-Neronian II, first recognized nearly twenty years ago in my doctoral thesis. Remember, the Neronians are those *Sapiens* populations who were exploring Europe a good ten millennia before what was thought. My post-Neronians are therefore the Neanderthal populations that replaced them in this same territory of the middle Rhône Valley. In any case, I am faced with an individual belonging to this population of the last Neanderthals. This final culture here closes nearly two hundred millennia of Neanderthal experience in the expanses of Eurasia. One of the last Neanderthals. It's superb. It's incredible. But we still have to work. Among countless questions, this one arises: is what we have here *one* of the last Neanderthals, or *the last* Neanderthal?

Luckily, my team is made up of about forty international researchers. The best in their disciplines. The Palaeolithic dream team. I'm going to check things out with my old pal Tom. Tom Higham, director of the dating lab at Oxford University, is one of the best daters in the world.

The last Neanderthal!

But, in order to date, you have to destroy, reduce to powder a certain quantity of bones from our Neanderthal. It's always difficult to bring yourself to make white powder from the last vestiges of an individual who died tens of millennia ago and whose remains have managed to cross the ages to reach us. Especially since the result is a matter of chance. These fossils are old and physiochemical or biomolecular analyses, all destructive, may give absolutely no results. It's a gamble, an assessment of the benefit/risk ratio. We might reduce our individual to powder for nothing – simply to be informed that it's all too damaged to obtain the slightest reliable measurement. But much worse than that, we can obtain measurements that may *perhaps* be valid. And in this *perhaps* lurks the doubt. Not the right doubt. Not the one that leads to questioning. But the one that blurs understanding. The one that obscures knowledge. The one that, once published, drowns the good results in a flood of uninteresting stuff. It then gets difficult to sort out the wheat from the chaff. As in the Middle Ages with the manufacture of counterfeit coins or devalued coins. When the market was flooded with silver coins but the silver was alloyed and diluted increasingly with base metals, lead or copper. Philip the Fair,[1] the counterfeiter king, would make the practice systematic, thus placing on the market many more silver coins than the physical silver in his possession. Hard cash, authentic coin of the realm, which has a vibrant chime when you throw it on the counter. Alloyed money sounds heavy, doesn't jingle. But people can buy the same products with it. Before long the reaction of ordinary folk is to withdraw real money from the market. And, as soon as the money starts to emit a false note when it jingles, the authentic coins, in real silver, become untraceable, archived under people's mattresses for later, when the bad money has been driven out. And so it is that, when the falsely jingling coins emerge, all the authentic coins disappear from the market. Bad money drives out good money. Seven hundred years after Philip the Fair, in science, things are quite similar. Good dates, fine analyses, robust, subtle results, opening up unexpected fields of thought, are commonly submerged in a broth of fragile data. And scientific data, the finding that changes the game, the discovery that could interest everyone, can be drowned in a flood of uncertain announcements. The discovery, the great discovery, the terrific advance, can be lost not only for the general public, but also for the biggest scientific journals, which are generalist by definition and may decide that a major discovery is no longer original enough to be presented in their columns. Good data see their value profoundly devalued. Bad money drives out good, here too. These realities affect all scientific disciplines and sorting out bad data can be considered a subjective approach that generally leaves room for fuzzy thinking. An open door, an avenue, to every kind of dominant thought. It's not my disillusion that I'm expressing here: these processes, these failings with which we have to deal, have already been widely dissected and are not specific to the human sciences – they affect all disciplines of thought and research, from medicine to astrophysics and quantum physics. We are here not faced so much with the question of an ethical fragility of science as that of the trial and error that build all science. And these hesitations, these uncertainties, are absolutely necessary for the exploration of knowledge. It's a blessing in disguise in a way: science also needs doubt, and time will sort it out. But we will remember from these limits of the explorations of the unknown that the advances of knowledge are fragile spaces and that the evaluation of their quality and their limits can be carried out only over the long term and only if it is freed from both dominant thoughts and immediate political or administrative issues. This need to protect research, and even more so the researchers who carry it out, against all forms of dominant thought brings us to the notion of 'research assessment' as carried out all over the world by those, researchers or decision-makers, who see themselves as the administrators of science. Apart from strict management – in other words a stranglehold on the development of thought, such an approach is both impossible and fundamentally dangerous in its applications. In the world of ancient dating, bad money is also widespread. It must be said that Neanderthal had the unfortunate idea of disappearing within the most uncertain temporal space, at the precise limit of our technological capacities to measure time using that remarkable tool, carbon 14. Disappearing somewhere before the fortieth millennium BCE, the creature is positioned precisely at that point in time where almost no team in the world can obtain reliable results using these dating methods.

Too bad: the creature died on the finish line. We could have easily dated dozens of sites to within a handful of millennia. But beyond the fortieth millennium, the portion of carbon 14 still residual in the bones has become so tiny that the slightest fraction of recent carbon, from the grease of the manipulator's hands or the smallest gob of spit, can tip the balance into a chronological elsewhere. So we purify the bone powder, we select, we separate the recent from the archaic. The Oxford laboratory has without a doubt the best team in the world when it comes to purifying samples and succeeding in dating the impossible. These procedures, able to establish a date on the basis of a single amino acid, are currently the most advanced in terms of our possibilities for purifying materials dated by carbon 14. No more pollution problems. The method they have just developed is already well proven and has made it possible to significantly bring forward the age of the last Neanderthal societies in Europe. From now on, we're no longer dating just the pollution affecting archaeological samples. Goodbye to pollution by recent materials: only ancient material is dated by carbon 14! We published this revision of the chronology of the Neanderthal extinction in the journal *Nature* in 2014, with forty-seven co-authors representing those in charge of all the European sites analysed. These dates included those of the Mandrin cave, at the entrance to which we've just found our Neanderthal.

Time for a phone call to Oxford to inform Tom of our discovery barely a year after the publication of this great overview of the last Neanderthals. Tom is enthusiastic. We are going to apply the 'single amino acid' method. We'll see.

However, there is no question of sacrificing the slightest fragment of these precious bones. Around our teeth lay tiny bone fragments, just a few millimetres long, that could well represent the remains of the mandible that disintegrated around these five teeth. But, well … It's quite impossible to tell a human splinter apart from a mammoth splinter and it would be a shame to give a date to a proboscidean while thinking we're putting a date to a Neanderthal. A few years earlier, during one of my stays in Oxford, I met a young researcher who was absolutely British, from the tip of his accent to the redness of his hair, and who in 2010 had developed a quite remarkable method. By analysing the components of a bone, he could recognize the species to which it belonged. The method worked with the same efficiency on a crumb of mouse bone as on an entire mammoth tusk. In the middle of a bag of thousands of millimetre-sized bone fragments, Mike could recognize a human vestige. A true scientific miracle, as if the method came straight out of a futuristic fiction. Now our millions of bone splinters, mainly picked up, on principle, for decades on our archaeological digs, were finally going to speak. As it happens, there are principles that produce results.

The strategy is simple: I obviously picked up the smallest millimetre of bone, including some tiny elements that appeared in contact with Thorin's teeth. The idea is that Mike will recognize, among these tiny bone fragments, those with a human signature before sending them to Oxford to be the subject of dating by the protocols developed by Tom. We will thus be certain that it is indeed our Neanderthal that we have dated. I select a handful of these small millimetre-sized fragments and send them to Mike at the University of Manchester. A few weeks later, bingo. Among the bones found in contact with the Neanderthal teeth, Mike has been able to spot four small fragments that have an unmistakably human signature. I ask Mike to forward this to Oxford. Thorin was discovered in August and we are coming up to the Christmas break but Tom has set the ball rolling and made the sample a very high priority. The sample will be processed ahead of hundreds of others from archaeological sites all over Eurasia, reducing our wait from two years to … a few months.

So it's May and Tom gets back to me. His message seems cheerful: the collagen in our bones was well preserved and his team has obtained some good results.

Dear Ludovic,
We've now completed the dating of Thorin's bone material. We have dated the hydroxyproline, which eliminates any likely contaminants. The radiocarbon measurement is 33,210 ± 780 BP. This result is obtained after correcting for

the background noise of the instrument. Obviously, this result is much more recent than what we expected based on the assigned level, which should be at least 43,000 years old considering that it's a Neanderthal. This would be the most recent Neanderthal ever reliably dated. Other Neanderthals just as recent have not been dated using such robust methods and when we redated them, their age was significantly increased, from 30,000 or 32,000 years to more than 45,000 years. The real age of the bones that we dated can be estimated after calibration between 35,000 and 39,000 years, with no possibility that the sample could be older.

In my opinion, the result is so surprising that we need to be sure of several things: (1) Is it really a Neanderthal? (2) Is the date reliable, and can it be tested? For this last question, the only thing to do would be to date the individual again. I think it's always better to confirm something so important with a double measurement!

Best regards,

Tom

The results are surprising. A few days later we phone Tom to clarify together the value of this dating. Tom's a cautious person. He has worked on most of the major sites in Eurasia and obtained results that systematically aged this singular moment of arrival of *Homo sapiens* on the European continent and the extinction of Neanderthal populations. His studies would shift the age of all Neanderthal sites thought to be less than forty thousand years old to over forty-two to forty-five thousand years old. Before Tom set up this vast programme for redating all the archaeological sequences on the continent, note had been taken of the persistence of Neanderthal populations in certain mountainous regions of the Balkans and the Caucasus and in the extreme south of the Iberian Peninsula – regions then considered as possible areas where Neanderthals took refuge and persisted for millennia after their extinction from other European territories. The methods of purification and the closely targeted dating of the Oxford teams had considerably shifted the time of this extinction. Now the problem was reversed: it was these same methods that made it possible to envisage a much later persistence of Neanderthals, somewhere around the thirty-seventh millennium, a good five millennia after the extinction of these populations on the European continent. On the phone, Tom told me that he wanted to establish a new dating of our Neanderthal, indicating that he would be surprised if this second phase of dating could fundamentally change the results obtained as the method used was currently the most robust in the world. But we both want to try to replicate this date. For one thing, I don't have any indication at the Mandrin Cave that would allow me to document any camps from the thirty-seventh millennium. So I can't link Thorin to anything known in my archaeological records. My last occupations from the Palaeolithic, well defined by nearly thirty years of research, were dated to forty-two to forty-four millennia ago. So how are we to explain that a Neanderthal body was abandoned at the entrance to the cave thirty-seven thousand years ago, especially since, statistically, the age of this body could just as well be only thirty-five millennia? For another thing, the main question was whether these human remains were indeed those of a Neanderthal, because we are not here dating just the survival of Neanderthal cultural traditions, but quite directly the remains of an individual who died at the entrance to our cave.

However, since sending these bone samples to Manchester, several months had passed and another sample analysed by the University of Copenhagen had allowed us to collect genetic information on our individual. I did not yet have the details of these analyses but Copenhagen had quickly contacted me to inform me that they had managed to extract ancient DNA from our individual, who was, without the slightest doubt, a Neanderthal. Not even a half-breed, as the first genetic analyses showed the absence of any genetic exchange between *Sapiens* and Thorin's population. Pure, authentic Neanderthal. So we did have a Neanderthal body, and with much more recent dating than all those obtained across Europe with the same carbon-14 methods. Thorin was far too recent. Too recent by five to seven millennia. Europe after the fortieth millennium had nothing in common with that of the previous three or four hundred millennia. Everything had suddenly changed and nothing seemed to have resisted this great wave, the new world that would reconfigure not only the European continent but the whole balance of human populations across Eurasia.

The imaginary realms lying between two worlds

In fact, whether Thorin is thirty-five or thirty-nine thousand years old – the two statistical margins of the dating of his remains – actually

48

changes little. After the fortieth millennium, there is no trace of Neanderthals or the Mousterian, the craft traditions associated with these populations, on the European continent. The vast dating program led by the Oxford laboratory in recent years made it unlikely that these populations had survived beyond this crucial fortieth millennium – for whatever reasons. And, to tell the truth, the reasons aren't really documented for us. Well, to be precise, they're not documented *at all*. It's as if the creature had evaporated. If we approach the question by trying to follow the last Neanderthal traces from site to site, we simply see nothing. Nothing but a before and an after, with no demarcation between the two. Without any upheaval of the biotopes. Without genocide. Without anything. Neanderthal, come out, come out, wherever you are! None answer. The Neanderthal has disappeared without the territories being abandoned. There is no vast European space without humanity. *Sapiens* arrived. First the one species then the other. And that's that. So this body, whatever its precise age, is a beautiful discovery. Usually, we don't find our old creature. We follow its paths, its traces, its flints, the bits of bison or mammoth bones that it cut up and all the things it left behind. The creature was not one to tidy its room and if you put a flint in the corner of a cave and came back forty millennia later it should still be where you left it. But what story does the object tell us – when we don't have the human being? Imaginaries. Rorschach images that pile up through time. Uncertain constructions; one more, one less. And that's that. The creature has gone. It has bowed out. This immense effort to redate the last Neanderthal sites led by the University of Oxford led to a first collapse of the interpretative edifice. Until then, the persistence of these populations could be envisaged in different areas of refuge, in geographical corners far from the major zones of exchange. As if the creature had persisted far from the world, in those mountains on the edge of Europe or at both ends of the continent, somewhere towards the southern tip of the continent, towards the rock of Gibraltar, or even in the frozen margins of the continent, beyond the polar circle, on the slopes of the Urals, that long mountain range that traces a very hypothetical demarcation line between Europe and Siberia. In a very hypothetical way, because although this low range of mountains stretches over two thousand kilometres, it never reaches two thousand metres in altitude. Its central part looks more like a set of plateaus than a mountain range properly speaking. Here the Chasovaya River crosses the ranges to join Siberia, mixing Asia and Europe together between two industrious cities, Perm the European and Yekaterinburg the Asian. But are these cities European or Asian? They're located precisely between two worlds. And it's this in-between-worlds that really interested me when, in 2008, I embarked on archaeological excavations along the Chasovaya with Russian colleagues from the Universities of Syktyvkar in the far north of Europe and from Perm in the Urals. My colleague Pavel Pavlov had spent his career identifying the Palaeolithic settlements of this vast Urals massif. Two astonishing sites had caught my attention, Byzovaya, on the slopes of the Polar Urals, and Zaozerye, much further south, in the subarctic zone but in the strategic area where Europe and Asia finally embrace along a vast river. From the height of its 28,500 years Byzovaya astounded me. These flint crafts were characteristic of the Neanderthal traditions that I know so well. But at least ten millennia after the supposed extinction of these populations. Here, the redistribution of the maps resulting from the Oxford programme did not apply. The age of Byzovaya was indisputable. The mammoth carcasses had remained frozen in permafrost. The quality of preservation of the bones was exceptional and all the indicators used by the daters were positive. As Tom Higham, a true exterminator of dates that were too recent, told me when I asked him his position on the datings we had obtained at Byzovaya: 'you know, Ludovic, all the indicators show that the preservation of the bones on this site is quite exceptional. There's nothing wrong with these datings. Byzovaya is probably the best dated Mousterian site in the world.' But there you have it, the Mousterian and all these old Neanderthal traditions were supposed to have disappeared a long time ago. There was no doubt, either, about the technologies present; they had nothing in common with the *Sapiens* traditions spread at the same time all across Europe. But was it possible to follow the trail of these popula-tions, not in the distant southern latitudes of old Europe, but locally, along the slopes of the Urals? Here, ancient sites are rare. Pavel Pavlov had spotted an interesting group alongside the last Ural foothills before Siberia: a riverside site where horse hunters had stopped a few millennia before our polar Mousterian populations. Zaozerye, however, was located five hundred kilometres further south than Byzovaya, in a very different environment, a natural passage between Europe and Asia, rather than in a polar corner of the European continent. The technologies of Zaozerye, about thirty-five millennia old, resembled all the other *Sapiens* technologies of the European Palaeolithic, with ornaments of shells or pierced fossils, bone awls and productions of really fine, slender flint, blades and bladelets, mass-produced. A classic of the old continental *Sapiens* industries, where Byzovaya gave us broad retouched cutting edges and familiar Levallois technologies in the Neanderthal style. Neither our excavations on the polar circle, however, nor our research in the heart of the Urals produced the slightest trace of human bones. Any interpretation thus relies solely on crafts, technological traditions and what they are associated with elsewhere, when we find comparable traditions ultimately associated with a small tooth or a fragment of DNA. But, in these distant fringes between two worlds, comparisons have to be made from afar. There, between our two sites, we make a leap of five hundred kilometres and five millennia. A mere blink on the scale of the Palaeolithic ages, certainly, but a vast universe when measured by our small human scales, and it is these human scales that are the only ones that interest us when we talk about the extinction of humanity. All the rest, all those immense mechanisms on a continental scale do not speak to us of humanity but of tables, trends, statistics, constructions. Our sciences. Our imaginations. Our Rorschach images? Well may we fear that the more these constructions take on the trappings of science, of very hard, quantified science, the science that offers neat and tidy tabulations, the more they move away from the actual matter that comes out of the ground and the more fragile, constructed and distant they are – I mean, far from the realities of the humanities. We must use all these sciences, extract the best from our best disciplines and force ourselves to keep our fingers dirty, with sand and clay still stuck under our nails. There is no science without this awareness. Without this proximity to the subject. Another form of participant observation, such as can be used in ethnography in a way. Keeping our hands dirty, as if making a demand, à la Sartre,[2] for a truth of shit and blood. Far from white rubber gloves. In this case, in Zaozerye, the hands were more dipped in ice and clay and in Byzovaya in sand and gravel. The beautiful objects seemed to point to Neanderthals in one case, *Sapiens* in the other and in this uncertain tangle there was not a bone to tell us who was doing what. But analogies exist – uncertain by definition. Between two worlds. Neither Europe nor Asia. Neither

Sapiens nor Neanderthal? Or both? Or not? Always the same doubt, whose surface I scratched in *The Naked Neanderthal*. Our naked imaginations. I know that this doubt is disturbing. It's painful, quite painful. So that's the itch you have to scratch without going as far as to draw blood. But sometimes it itches too much. You see?

In any case, the dating of Thorin, my last Neanderthal, pointed clearly to that space–time where Neanderthal was supposed to have left the scene some time ago already. The astonishing polar Mousterian of Byzovaya was even a little more recent. The beads of Zaozerye were precisely the same age as Thorin. But how? How, even on the edges of the polar zones, could we already be seeing *Sapiens* ways of life flourishing while in one of the largest migratory corridors of the European continent, on the banks of the Rhône our Neanderthals apparently continued their old ways? As if nothing had happened. As if, here, the Neanderthals didn't give a damn about European history in its entirety. About its migrations and all its new ways of thinking about the world. About understanding it in the same way that we did. As if the collapse had never happened.

In any case, it was from this science, this hard and quantifying science, that totally contradictory results would emerge. So contradictory that it would take us more than four years to begin to see things more clearly. Here, a battle would be joined between the human sciences and the hard sciences. A battle of field archaeology against physiochemical measurements and against palaeogenetics combined. Not a fight to the death but a good-natured game of chess because, in science, data that do not agree should always be welcome, even if this last proposition fits so badly with the nature of *Sapiens* as I have just presented it. But this game of chess, good-natured as it is, is quite fascinating since it sets up totally contradictory, incompatible results, between the interpretations resulting from my archaeological fieldwork, the carbon-14 measurements and the data resulting from the DNA reading of my Neanderthal.

Upside down … Thorin is one hundred and five millennia old!

Dear Ludovic,

We now have good palaeogenetic results and are able to propose a chronology for Thorin. This individual is part of certain ancient Neanderthal populations recognized in Europe and our statistical models now allow us to calculate an age for this individual. Thorin is, according to our calculations, approximately 105,000 years old and statistically speaking cannot belong to a population less than 100,000 years old.

I am very curious to know what you think about this, and how these results agree with your archaeological data.

Best regards,

Martin

Martin is one of those excellent palaeogeneticists from the University of Copenhagen. One of the best teams in the world. And for palaeo-genetics, Thorin is indisputably one hundred and five millennia old. Oxford offered us an age of thirty-five millennia, while Copenhagen shifts our Neanderthal back to a period fully three times older. We're shifting into *something else*. A quite uncertain *something else*. For me, an archaeologist, if we base our conclusions on the stratigraphy of the cave, a 35th-millennium Thorin is 7,000 to 15,000 years too recent. If he is 105 millennia old, he is too old by 50 to 60 millennia. And there's no cautious approach that allows us to position ourselves halfway. The data from archaeology lead me to reject both the radiocarbon data and the genetic data. And the three disciplines contradict each other harmoniously. It all feels uncomfortable, it grates – so there's something interesting going on. Something to understand. These are interesting situations because they will have to draw on a set of data from other disciplines to try and sketch out a new interpretative framework. By the 35th millennium, Neanderthals must have long since bowed out definitively, but a general interpretative scheme of the data from conti-nental Europe cannot exclude the existence of infinitely more complex processes in which populations could have persisted much longer. But such an interpretation seems very unlikely. The Rhône is the opposite of a peripheral zone where archaic societies could persist for millennia, out of step with time and the great march of history. The Rhône is the central axis of continental movements, the junction of continental masses moving towards the Mediterranean. In the thirty-fifth millennium, a few kilometres away, one of the oldest and least expected decorated caves in the world was recorded: the astonishing Chauvet cave. One of the most remarkable too. It is in this space of circulation that we record the oldest *Sapiens* migration towards the European continent, and the conjunction of this very old population and these very ancient expressions of wall art is far from being a chance encounter for me. I would even say that the Mandrin cave provides the illumination, the thickness of time, which allows us to understand the Chauvet cave, and to give it a context. Not only in showing how very ancient migrations of modern populations into this territory took place but also because the Mandrin cave contains brief human settlements, around the forty-second millennium, occupied by other *Sapiens*. If Chauvet was occupied by the Aurignacians, one of those old *Sapiens* traditions, Mandrin was occupied seven millennia earlier by what are known as Protoaurignacians, direct ancestors of these same traditions. An amusing anecdote is that in the Mandrin cave we found a fragment of the ancient vault of the cave that had fallen into our archaeological levels – one of the many fragments that usually reveal soot deposits. But this time the fragment was not tinted with the natural black of soot, but red … This fragment shows us that the wall was painted red, locally covered with ochre. We have the recording of parietal expressions of which we know nothing. Typical of cave art, probably, but far from the bowels of the earth, in a small cave that is an open mouth on the world, rather than an intimate, underground environment. Mandrin is a shelter under a vault, a bubble in the rock overlooking the immense Rhône furrow, not a closed space, hidden in the bowels of the earth, difficult of access, concealed. Here everything is open to the full light of day. And these Protoaurignacian settlements in Mandrin are rather trivial – a few tools, a few weapons, that's all. The analysis of the soot shows that this settlement is one of the most discrete in our entire archaeological sequence, over 80 millennia. As if the cave that suited the Neanderthals perfectly did not suit these populations, in their organization. These very discrete passages of *Sapiens* of the forty-second millennium give the impression that the place had been taken over. Their settlements do not express themselves again – as had happened twelve millennia before them with the Neronians – until a few seasons after the last Neanderthal camps in the cave. And despite the discretion of these settlements, the *Sapiens* populations deposited ochre on the walls. Signs? Symbols? We don't know; the fragment of the vault measures just a few centimetres. But the wall was painted. Are they the signature of a symbolic takeover of the place? They at least sign a certain way of being in the world. I pass and you pre-pass, in a way. Could Neanderthals still have persisted here seven millennia after these *Sapiens* repainted the apartment? It's possible, after all. There's nothing linear about historical processes or human societies. Nothing reducible to any form of statistics. Nothing predictable. But we would then have the longest persistence of these Neanderthal populations on the European continent.

However, another problem now arose for me. For my geneticists, there's no doubt about it, my Neanderthal is one hundred and five millennia old. At best one hundred millennia old. No less. Here I'm faced with a more thorny problem because, while I can try to construct a model to explain a Neanderthal who is too recent, I can't construct a serious model that would allow me to understand how these hundred and fifth-millennium populations managed to send me a body from an archaeological level that would not exist until sixty millennia later ... That's a trickier proposition, as you will agree. The general principle of archaeology is that the more you dig, the further down you go, the further back in time you go. And if we do have levels of one hundred and five thousand years in the Mandrin cave, in this zone they're located a whole metre under the body of our Neanderthal. And in Mandrin everything's frozen, nothing moves, objects from the one hundred and fifth millennium remain in the levels of the one hundred and fifth millennium and those from the forty-second remain in those of the forty-second. Dozens of radiometric datings confirm the remarkable integrity of the archaeological levels. The flint objects themselves are very different from one level to another, both in the technologies used by the different populations that occupied the cave and in the choice of rocks used. We thus have levels composed of black flints that cover levels composed of blonde flints without any of these blonde flints being found in the same level as the black flints. And we're talking here about tens of thousands of objects. The anomaly here is as visible as vinegar in oil. So I can't imagine a body abandoned in a stratum from the one hundred and fifth millennium being found in a stratum from the forty-second, crossing several geological layers and several very different archaeological levels, without leaving any trace of this passage through time. No, this body hasn't moved. It must have been abandoned where we found it. But how, then, can it be so old? And my geneticists are categorical on the matter. We can imagine a scenario in which a recent body is found in ancient archaeological levels. If the group decides, for example, to bury the body of one of their dead:

they dig the ground and deposit the mortal remains in ancient strata. But the practice will leave obvious traces of digging, turning over soils and objects. A hole, a pit, even a small earthworm gallery, can be easily spotted and followed during the excavation. And, above all, the proposition can't be reversed. While I can deposit a body in an older stratum I cannot deposit a body in a future stratum – one that doesn't exist yet and will be deposited somewhere higher up, one or more metres above the ground on which I am currently walking. A body cannot be abandoned hanging in the air. If aborigines buried the bodies of their dead at some height in the hollowed-out trunks of very old trees, no matter how long the tree lived, in a hundred years, or a thousand years, the tree would die, dry out and end up turning to dust on the ground. My best model can therefore offer this body a respite of one thousand to one thousand five hundred years, corresponding to the age of the oldest tree that could persist in our latitudes, an old oak or an old yew, for example. But over 60 millennia, I am still short of at least 58,500 years. So I don't have an explanatory schema that would accommodate the analyses of palaeogenetics. Ultimately, the most original schema that I can attempt to construct would be that of a body deposited in a recess in the rock. Thorin's body is indeed laid out along a bend in the wall but it's in the sandy sediment, not placed on the rock. However, we know of many populations who bury the bodies of their dead in the hollows of rocks, on stretches of cliffs. Among the Dogon, in southern Mali, the smallest corner of the cliff is filled with bodies. I was able to see the remains wrapped in cloths, then hoisted several dozen metres above the ground in the nooks and crannies of the Bandiagara cliffs overlooking the villages. The climb to the burial sites is sometimes arduous, but one can see how these piles of bodies accumulate through the ages, the decomposition of the oldest bodies freeing up space for the new dead to be piled up. If I imagined a Bandiagara scenario in Mandrin, it would be necessary for the Neanderthals of the one hundred and fifth millennium to have deposited this body in a corner of the rock, overhanging at eye level and for this body to have remained like this, overhanging, before slipping into the sediments at the precise moment when these sediments, after having gradually settled in the cave, reached the level of the overhang where this body had lain for sixty millennia. Two further conditions would have been necessary for this, the first being that these remains be protected from carnivores – lions, panthers, hyenas, or wolves – attracted by the smell of rotting flesh; and second, and more difficult, these bones would have to have remained anatomically connected to each other in this corner of the rock and then, sixty millennia later, slide into the ground – not bone by bone, gradually, but rather en masse, so that the teeth maintained their respective positions relative to each other. It's not impossible. But it's incredibly improbable. It begs a lot of questions. But in humans, it's true, the improbable always lies in wait. It is difficult to exclude, a priori, a succession of improbable movements that made an astonishing scenario possible. Nothing allowed me to reject a Bandiagara scenario. Chance would then have led these remains to slide, in one block, tens of thousands of years later, into the ground near this wall – precisely where we found them. I thought Thorin was forty-five to fifty millennia old. He was one hundred and five millennia old. You have to be able to rethink your ideas. Accept the improbable. But all this is so uncertain. And yet, my geneticists seem very sure of themselves. Which discipline, genetics or archaeology, the latter a fragile human science that looks at and interprets primarily just with our senses, our eyes, and our hands, here offers the correct interpretation of what actually happened? Really, I didn't expect it. But we have to accept the improbable.

Dear Martin,
A big thank you for your message. To be honest, I don't really have an archaeological context that would allow me to explain how a body dating back 105,000 years could be found here in strata that are between 42,000 and 50,000 years old. The only possibility I can imagine would be that the body was hidden for 60,000 years in a crevice in the rock very close to where we found it, before sliding into the levels corresponding to our settlements of the last Neanderthals, but this interpretation here seems very fragile to me. We'll have to push the investigation to try to understand what's happening.

A big thank you for the work of your team. We will try on our side to deepen our analyses, to see if this individual can fit into these old chronologies. Best regards,
Ludovic

If Thorin is one hundred and five millennia old, he lived in a temperate, forest environment. Before the last Ice Age. These old settlements of 'warm Neanderthals' are well recognized in the Mandrin cave. I call them the peoples of the forest – very distinct from the peoples of the great cold prairies, populated by horses and bison, from forty-five millennia ago. If Thorin is indeed one of the representatives of the people of the forest, he experienced a climate very close to the current climate, perhaps even more temperate. One solution emerges fairly quickly within the team. If he lived in these forest, warm environments, his bones, and in particular his dental enamel, must have recorded markers specific to these climatic environments. The dentine could be too porous and subject to too recent contamination but a tiny fragment of enamel should provide us with a remarkable array of information on this individual's life and environment. The analyses are almost non-destructive, and a fragment of three square millimetres can perfectly well be the object of this research. Luckily, a very small fragment of enamel had been found in the ground, detached under one of Thorin's molars, and the tiny sample had been sent to the University of Tübingen in Germany in order to analyse its isotopic markers. These indicators are characteristic of the environment in which a living organism lives and can change with consumption, climate and the environment in the broad sense in which this organism lives or has lived. According to these indicators we could know if Thorin lived in a temperate, wooded environment or in a cold environment of grassy steppes. It would also be possible to assess whether Thorin had grown up and lived in the limestone geological environment in which his remains were found. Four families of isotopes would be sought and then compared to the same markers recognized through analysis of other Neanderthal remains in Mediterranean and Atlantic France, and as far as Belgium. These measurements would then be compared to those from bison and ibex remains found not far from Thorin's body. And to perfectly calibrate these studies other human remains only four to five millennia old would be analysed in order to assess local isotopic markers in temperate periods. We should immediately be able to ascertain whether our Neanderthal had also lived in temperate climatic environ-ments. This tiny fragment of dental enamel should reveal to us the landscapes in which Thorin lived.

A long year would pass before Hervé Bocherens, master of these isotopic analyses, could build up a first fresco of the environments and climates recorded in this small dental remnant. One fine morning the results finally came in. The different isotopic markers all revealed an open environment without real forest cover, cold, comparable with those recognized on other late Neanderthals between forty and fifty thousand years old. There was no marker providing the slightest indicator of a temperate environment characteristic of the recent human remains that we analysed in parallel.

But these indications are not sufficient. Thorin could still be one hundred and five thousand years old. Let us imagine that he belongs to this distant temperate phase before the great glaciations but that he spent most of his life in a mountain environment – that he died in Mandrin but was in reality a mountain man, a Neanderthal from the alpine environments, much more hostile climatically, or a man from beyond the western banks of the Rhône towards the Massif Central. There, a harsh climate still whips the high plateaus and the mildness of the Mediterranean, despite being so close, soon becomes just a vague memory. In these valleys and on these high plateaus a few weeks of stifling summer are succeeded by interminable months of an endless winter. It's the whip of the winds, the cold of the burle,[3] the fearsome breath of the high plateaus on the sparse vegetation. If Thorin is a man of these plateaus, or of these Alpine cliffs, our cold markers would not tell us about the time before or after the glaciation but simply about a geography and a more or less mild, more or less hostile environment. More or less hostile. Yesterday, as today, there is little reason to die where one has lived. A man moves around. Especially a nomad – a nomad doesn't stay still. Not even a Neanderthal. While Neanderthals are not known for being great travellers, I had been able to identify, in some archaeological collections of the Loire Valley, movements of flint over more than three hundred kilometres – but three hundred kilometres towards the northwest and three hundred kilometres towards the northeast and three hundred kilometres towards the southeast: remarkable circulations that could, on a single site, affect more than a third of the current territory of France. Populations move, objects are exchanged, information and genes pass from group to group. Some of these objects showed movements between the two ends of the Massif Central, from north to south. If Thorin was one of these men of the high plateaus, or had spent a good part of his existence in these environments, my decoding stratagems were useless.

If he comes from those frozen plateaus, if he hung around in those sunless valleys where, until the nineteenth century, they built stone igloos to sell very hard ice cream in the city bistros in the summer, my isotopes no longer tell me about climatic phases, no longer position me in time but in moments and anecdotes specific to the life of this man. My isotopes would be nothing more than an illusion. Other indicators had to be evaluated. Some markers still need to be analysed and combined. These cliffs, these harsh, high plateaus can also be identified through isotopes. These Auvergne landscapes are composed of crystalline massifs, granites, gneisses, schists. The rivers that cross them carry all these components with them and when humans or animals drink this water, these crystalline indicators accumulate in their bodies, in their bones, in their teeth. But Thorin's analysis also concerns the search for crystalline indicators. And Thorin doesn't show any of these traces. Nothing allows us to suspect that he ever lived in these particular environments.

From rivers that unite to rivers that separate

Even more interesting, these markers are carried by, and are present in almost all of the rivers of the Ardèche – and Hervé is very surprised. A few years ago he analysed a Neanderthal tooth found a stone's throw from the Mandrin cave about twenty kilometres away but on the opposite bank of the Rhône, in Ardèche. This Neanderthal from Ardèche who, like Thorin, lived on the banks of the Rhône in a very similar environment, has very clear crystalline indicators. Thorin has none. The two individuals lived in similar biotopes but were found on the eastern and western banks of the Rhône respectively. The river here seems to distinguish two realities that impact the populations on both banks in different ways. Hervé and I call each other at length to discuss this perplexing discovery. Indeed, our archaeological knowledge allows us to distinguish very clearly Neanderthal and *Sapiens* mobility, with incomparably larger territories among our *Sapiens* ancestors. However, Neanderthals do not resemble frozen populations. A few years ago, my work on the banks of the Loire showed precisely that it was not so much the territories that were less extensive as the circulation strategies that distinguished these two populations. If *Sapiens* did not hesitate to carry large quantities of very good flints over hundreds of kilometres, Neanderthals travelled on the same territories while keeping only a few flint tools in their pockets, in addition to what our nomads knew they would find in the different terroirs crossed. It was not so much mobility or territories that seemed to distinguish our two humanities as certain ways of registering their presence. Here the maps showing the circulation of *Sapiens* and Neanderthals overlapped very clearly but illustrated very different mobility strategies. As such, these divergences could be due to simple habits, ways of travelling. However, it was not possible to determine whether these flint objects really travelled in the pockets of their artisans or whether they circulated from one place to another, through exchanges, when different groups crossed paths during their seasonal movements. Perhaps it was a bit of both. This site on the banks of the Loire revealed rocks that not only came from very distant origins whose sources could be recognised over three hundred kilometres away but also came from opposite directions. The cumulative distances then seem so vast that they hardly reflect the seasonal cycle of movement of a single population. Here, on the banks of the Loire, it was probably peoples from the Paris Basin, peoples from Burgundy and peoples from the Ardèche who must have met at a very particular time of year to hunt vast herds of horses on their migration route. It was an opportunity to meet up, to exchange, to talk, to feast, like Inuit populations separated for too long by the lengthy winter who meet up as soon as the weather improves to consume together the last, now unnecessary reserves of winter. But these flint objects from distant horizons were almost invisible archaeologically. A dozen objects, no more, lost in a collection of nearly ninety thousand flints. The remarkable find was almost imperceptible. But on opening the sieve bags where the archaeologist places the millions of flints measuring just a few millimetres, I find other traces, just as imperceptible. Among these millions of remains, 568 small millimetre-sized flakes intrigue me. Sorting the bags and analysing these tiny elements would allow me to recognize that the dozen exotic flint points that I had found were not representative of what had happened here. I would be able to calculate the number of points that had been brought to the banks of the Loire, had been resharpened there, generating a handful of small splinters abandoned on the ground and had finally returned to the pockets of those who had used them here. And this study of the 93 invisible objects would reveal to me that 202 points, from very distant origins ranging from the Atlantic to the Mediterranean areas, had here passed through, along the banks of the Loire and had been used here; their cutting edges had been resharpened in this place but the objects had left with their Neanderthal artisans. These ghostly objects were at the same time very present. A handful of objects that quite profoundly modified one's potential perception of certain strategies of circulation among these populations. The objects brought and taken away from and to such distant geographical origins suggested that it was indeed the Neanderthals and not only their objects that, in a series of exchanges, circulated here over very vast territories. Like *Sapiens*, these nomads also had itchy feet.

If I did not see such circulations in the Rhône Valley, the river systematically appeared at the heart of the territory of these Neanderthal populations, whatever their traditions, for a very long time and in all climates. One hundred and twenty millennia ago, in a warm context, the forest peoples went to look for many, and sometimes all, of the rocks needed to shape their weapons and tools, on the opposite bank of the Rhône, commonly within a radius of twenty to forty kilometres. These methods were still the same sixty millennia later. Of course, the same resources were not used, nor in the same proportions, nor in the same way. But, within these societies that were culturally distinct and also lived in radically different climatic environments, one point in common could be identified. The Rhône was always at the centre of the territories. In the Neanderthal settlements on both banks of the river one could recognize the working of a significant portion of rocks, sometimes most of them, from the opposite bank of the Rhône. From the 120th to the 55th millennium the Rhône was the centre of trade and communication networks, whether in a temperate context, with peoples living in vast primary forests, or in the context of an Ice Age. Obviously, in these polar environments the great river must have been frozen by ice for nearly eight months a year. The river then did not exist as a geographical line physically distinguishing territories but represented a vast frozen white line. A highway naturally positioned at the location of our own autoroutes. We have merely highlighted with bitumen a major natural migratory axis on a continental scale. But the experience of this geographical reality was radically different for the peoples of the forest who were confronted with a vast expanse of water already spewing its one thousand cubic metres per second.

Even if the river could spread out into many branches, crossing such a river could not be an easy matter. And yet, even in this unique temperate context populations regularly crossed the great river that already lay at the heart of their territory. And this reality, across tens of millennia, in all climates and in all human societies whatever their traditions, would suddenly stop after the fifty-fourth millennium. From the peoples of the forest to the first *Sapiens* migrations, trade flourished. The Rhône remained a space for circulation and exchange. But after this *Sapiens* parenthesis human populations would never again cross the Rhône. The reasons for such a sudden, sharp split are completely unknown to us. There's not the slightest trace of any crossing of the great river. Not the slightest archaeological evidence. The situation is astonishing, almost gobsmacking. It's anachronistic compared to the previous sixty millennia. This unique situation would last twelve millennia. During this immense thickness of time, the last Neanderthal societies seem never to have crossed the great river again. We can recognize two major, clearly differentiated cultural phases. The first Neanderthal peoples to reoccupy this Rhône area seem to have come from the alpine areas of the Diois region, more than seventy kilometres northeast of the Mandrin cave. We know that in this first phase of reoccupation of the territory, after the *Sapiens* parenthesis of the Neronian, these societies were indeed Neanderthal. I have found two indisputably Neanderthal children's teeth in these settlements. This split in territories would also be evident during the following cultural phase, from the 50th millennium until the Neanderthal extinction eight millennia later. It was within these archaeological levels that I found the body of one of the members of this population, Thorin. During these settlements, representing the final cultural pulsation of the Neanderthals, the territories seemed to stretch along the Rhône. But they did not cross the river. They no longer crossed it. However, the climatic context was very harsh. The environmental conditions were polar. The great river must nevertheless have been an immense white highway frozen by ice for a large part of the year. And no one crossed it? But the river could not be a physical border, unless you refuse the idea of skating over the frozen expanses. It is in this context that I proposed, in 2004, that this border, this obvious, all-too-obvious border, could not be a physical border. It was necessarily a social border. The groups on the opposite bank of the Rhône no longer allowed access to their territory. This was the expression of what would perhaps be the first, the oldest human border ever drawn. Among traditional populations, as on our globalized planet, as we are beginning to see again, distances are never kilometres. Distances are always social. If my relations with the populations located on the opposite bank of the Rhône are good, the social distance is small. If they're bad, the distance is immense. The question, the only question, is always that of the definition of my relations with my neighbours. And, since this astonishing frozen border cannot be physical, it must be social. For some unknown reason, the peoples located on the right bank and the peoples located on the left bank admonished each other from afar. They no longer traded – not at all, after sixty millennia of circulation and union by water. And it is in this astonishing situation, unprecedented in the history of these societies, that we are going to experience, here as elsewhere, what is perhaps considered the greatest extinction of humanity of all time. In this very particular context of borders, social distances, of a breaking of all the links that had united populations for tens of millennia. It was in these words that I would pose these concepts in my doctorate almost twenty years ago. But it was a daring proposition. I projected human circulations, interactions and social constructions from the presence or the sole absence of circulations of objects, of materials. Real territories are invisible, even more so for the archaeologist who must accept this frustration, this permanent intellectual juggling between the invisible and the uncertain. In the shadows … always.

But this is indeed what I saw here. And what I proposed and defended. We could trace ancient social boundaries, gauge relationships and breaks in the traditional groups of this territory. And now, almost twenty years later, Hervé expressed his amazement to me. The isotopic markers of our individual are radically divergent from those of another Neanderthal located a few hours' walk from here, but on the opposite bank of the Rhône. Thorin, who is a middle-aged man, had never crossed the Rhône. Or if he had transgressed this limit of the great river, he had never drunk water from the opposite bank, in which case the crystalline markers carried by the Ardèche rivers would have been inscribed in his flesh and bones. They would be visible, discernible. This individual seemed to fit precisely into what I had recognized as the organization of the last Neanderthal societies. Precisely into the archaeological levels where this body had been found.

Could Thorin really be 105 millennia old when his bones reveal a battery of cold isotopic markers? The absence of crystalline indicators excluded the possibility that he had lived on the high plateaus of the Massif Central or in some other mountainous region. And yet this individual had spent his life in extremely cold environments. All of this information could not fit with the results obtained by the Copenhagen geneticists. *Nothing* fits in this story, in these measurements, in these quantifications. Thorin's isotopic analyses are not only consistent with the climatic environments of the archaeological level in which I found him, but his isolation from the western areas of the Rhône seems to place him very precisely in the very specific phase experienced by the last Neanderthal populations in the Mandrin cave.

We were faced with three possibilities: archaeology indicated that we were in the presence of a Neanderthal between 42,000 and 50,000 years old; radiocarbon placed him around the 35th millennium; and genetics sent him back to the one hundred and fifth millennium. These three paths led us in radically opposed interpretative directions, and no conceivable middle way that could slice, with sweet modesty, the pear in two somewhere between 105,000 and 35,000 years. Only one of these three disciplines was right. Or they were all three wrong. Who knows?

Light beyond the unknown lands

And the weeks that pass, then the months, the years, with pieces of the puzzle that slowly come together, agree with and contradict each other, hesitate and grope. As I told you, research is a step into the unknown. An uncertain step. There was no doubt that we were here in terra incognita, in a zone of basic research lying at the crossroads of disciplines that were all complex and, at this ultimate level of expertise, were mastered by very few specialists on earth. And here we were, groping, trying to align our views, our understandings, our measurements. To be honest, in science, we are generally happy to see our thoughts, our suggestions and our hypotheses overturned. And we also like to understand how each new advance rearranges the overall structure of our perceptions – how the elements of the system impact, influence and modify each other directly or indirectly.

And that morning, a new message came from Tom from Oxford. A new round of dating on other micro-bones of Thorin had just given divergent results. There was no longer any question of an isolated individual dating back to millennia after the extinction of all the other Neanderthals. Several measurements had just been obtained. They placed Thorin somewhere between the 45th and 50th millennia. Could some of the mists be starting to lift? That would still be too easy…

These new datings don't tell us that Thorin is between 45,000 and 50,000 years old. The measurements are compatible with this timeline, but only represent minimum ages. So we realize that Thorin might actually be 45 millennia old … or 105 … The measurement remains potentially open to older timelines. On the other hand, it eliminates one of the three paths. Thorin may actually be one of the last representatives of Neanderthal societies, but he does not prove the survival of these populations millennia after the extinction of all the Neanderthals in the old world. And the idea of the 35th millennium saddened me a little. If one must always be ready to invite the unthinkable, to welcome it with benevolence, I had neither archaeological context nor interpretative scheme that helped me to understand what I was now confronted with. My archaeological records here stop at the forty-second millennium and the few missing millennia meant I was unable to define the organization of Thorin's society. I had no flints for the 35th millennium, no bones that could structure an initial hypothesis as to the persistence not only of such societies but of these populations as a biological reality. Could Neanderthal groups have survived for millennia without leaving any trace either in this cave or in any other site in Eurasia? Could this long isolation have signalled a very slow extinction of the Neanderthals and their exclusion from the *Sapiens* societies that at that time occupied almost all of the planet's latitudes? How could one of the largest migratory corridors in the Mediterranean have provided a dwelling for the last representatives of a population that was extinct everywhere else?

My polar research had nevertheless confronted me a few years earlier with the possibility of such a scenario. I had then suggested the possibility that these populations had continued for several millennia beyond even the 35th millennium that Thorin's dating indicated to us. But there was a context, a logic, in the polar circle. The fringes of the Polar Urals are not the shores of the Mediterranean. However, in both these cases we were on the banks of a major river. Imagine that the Pechora River has a flow rate almost twice that of the Nile. But this polar area was situated far from all the obvious migratory axes without us yet being able to understand either the initial populations beyond the Arctic Circle or the precise organization of these first boreal peoples who seem to have invested the northernmost regions of the planet even beyond the 50th millennium. On the fringes of the Polar Urals, at Byzovaya, accumulations of mammoth carcasses have been found abandoned along the immense Pechora River. Around forty mammoths were put to use, being cut up and consumed. There were also mammoths found around Thorin as well as the woolly rhinoceros but our Mediterranean Neanderthals did not concentrate on these large species, preferring horses, bison, deer, or ibex. The exploitation of the mammoth was, rather, a boreal mania. The peoples of the mammoth were a sort of continental preamble to the Nordic whale peoples. And while, in the thirty-fifth millennium, the extinction of all Neanderthal populations seems to have been a fait accompli for a long time, doubts persist at both ends of Europe. Are we sure that certain populations could not have persisted on the fringes of their world, to the south as well as to the north? It's still impossible to firmly settle this question, but it's remarkable that the question remains unanswered on the furthest edges of Europe, either because the populations managed to persist in the distant margins of our continent, far from all the major natural routes of circulation, or because the old Neanderthal traditions, their ancestral knowledge and ways, were here able to survive for millennia after their abandonment throughout the old world. The legatees of these traditions could have been Neanderthals or *Sapiens*. And what was the place of the more or less interbred populations?

In any case, Thorin's question too was now as yet unanswered, no longer between three possibilities, but between two. The hypothesis of a late persistence has to be abandoned. At the heart of the migratory axes, such persistence would have been astonishing, and fascinating. But what precise historical processes caused the extinction of these populations, this body does not tell us. Two paths remain to be explored. Only one can be correct. Did he die 42,000 to 50,000 years ago – or 105 millennia ago? Palaeogenetics also told us that Thorin was a male and that he showed no genetic exchange with the *Sapiens* populations. This is an interesting context since, if archaeology was right versus genetics, he belonged precisely to the populations reoccupying this territory after the very first *Sapiens* settlements on the European continent. But whatever his age, this absence of genetic exchanges was in itself not very astonishing: no Neanderthal, whatever his or her chronology, shows the existence of such genetic exchanges between these populations. I already noted the implications of this remarkable paradox in *The Naked Neanderthal* since all ancient *Sapiens* actually show Neanderthal antecedents. Either Thorin was one more element fitting into my love–hate scenario or Thorin actually belonged to an ancient population over 100 millennia old, which in itself explained this absence of exchange between the two populations. But how can we decide? We now know that Thorin is a 'cold Neanderthal', one of those populations from the great grassy steppes, but how can we be sure that we've taken everything into account? Only obtaining direct dating could now allow us to move forward, but carbon 14 has shown its limits here, even with the best team of daters in the world. We need to implement other dating strategies.

Our discussions with Tom allow us to draw up a plan. We're going to turn to an Australian specialist who has developed certain remarkable methods based not on carbon but on the uranium series, another radio-active component that can be found in teeth. Renaud Joannes-Boyau is a researcher in Lismore, New South Wales, and has established a method for dating whose resolution equals that of carbon, with virtually no damage to the material studied. The dating is carried out on samples of just a few microns of enamel – extractions invisible to the naked eye. This method is of particular interest to us because it can be applied to relatively old objects where carbon constrains us to periods of less than fifty millennia. The result is nevertheless far from certain. The measurement may be negative, or provide no precise result, or offer no robust dating.

If we fail to date Thorin directly, we will just have to wait and see, since the methods of archaeology and palaeogenetics lie in two opposing areas, without any objective possibility of resolving our interpretative differences. However, in this game of chess between the humanities and the hard sciences, the two disciplines are on unequal ground. Each one will be able to present its argument, explain how our results diverge and why one thinks that Thorin is less than fifty thousand years old and the other is certain that he is more than one hundred thousand years old. This is generally the moment when everyone throws in the towel. The subtle, interpretative humanities, which are based above all on words and minds, get pushed aside by the hard quantified sciences. Who is ready to follow a thinker rather than the analysis of nucleic acid sequences? We'll see.

We are sending some of the teeth of the wild animals found around Thorin to calibrate the method and see if the measurements obtained on Thorin's dental enamel are reproducible on the remains around him. We should have the results fairly quickly…

But that would be too good to be true. For at this moment, the borders close in doubt, amazement and fear. The time of Covid has arrived. Everyone goes into and comes out of lockdown at their own pace. And, of course, at just this moment the laser in Renaud's measuring instruments breaks down. The laser engineer will find that the problem was related to mirror blistering due to humidity. The part has to be replaced and shipped from the United States. But in the strange phase of Covid the magic mirror will take an interminable time to reach us. On 8 October 2020 Renaud finally gets back to me:

Dear Ludovic,

I finally got a measurement of 43,000 years for Thorin. I'm convinced that the age of the fossil is very likely close to this. I'll make some more measurements tomorrow while trying to map the samples in order to obtain some very reliable arguments for this sample. It's very unlikely, I'd say almost impossible, that Thorin is 100,000 years old or even 80,000 years old, to be honest. I'll send you my report early next week.

All good wishes,

Renaud

This date, 43,000 years, is not only unlikely to be significantly modified but the animal bones around Thorin, remains of the fauna hunted by Thorin himself or those of his group, subsequently delivered precisely the same measurements. These measurements are very closely connected to the ages obtained concerning the archaeological level in which Thorin was found. Finally, the batteries of analyses carried out on these bones make it possible to exclude any process allowing us to consider that they're fundamentally older. Our two interpretative paths finally seem to converge. Thorin can only be between 43 and 50 millennia old. He would indeed constitute, as the archaeological data indicate, one of the very last representatives of Neanderthal populations. Thorin does indeed belong to those groups who, for an unknown reason, but related to their social rules, their history, their organizations and, for quite mysterious reasons, will never again cross the Rhône.

But how is this possible? The palaeogenetic teams seemed absolutely sure of their results. Could the molecular clocks from genetics really have been wrong? I'm very curious to know what Martin will tell me about our different results because there are now three disciplines – archaeology, isotopes and uranium series – that indicate independently of each other that Thorin must be one of the representatives of the very last Neanderthal societies.

I forward all of our results to Martin, who is indeed as surprised as he is interested by our conclusions.

Thank you, Ludovic, this is really very interesting. It certainly raises a puzzle regarding the approach to molecular dating.

We will need to repeat all of our analyses but now add Thorin as a time calibration point going back to the 40th–50th millennium, and then re-estimate all the ages of the other Neanderthals known to date through all of our statistical models.

From this, we should be able to see how a recent age for Thorin would affect the dates of the other Neanderthals and assess whether these results are consistent and logical. If so, we will have to speculate on possible explanations (is it related to differences in generation time, age at reproduction in Neanderthals? …), but first let's see what it looks like.

I apologize for the length of our many email exchanges, it's sometimes good to write things down to clear things up in one's mind :) If all this seems reasonable to you, we'll get back to you very soon with the remaining results, so that we can discuss the implications of all this together, and the different possible interpretations.

I can't wait to see the results of these new analyses in detail.

Martin

Something is clearly eluding us. Martin will relaunch many statistical models to position each ancient population in relation to the others and establish phylogenetic trees, graphs that draw the relationships, crossovers, divergences and isolations between human populations over time. By cross-referencing these data with the age of the sites from which these genetic data were collected, it is possible, for example, to determine the chronology of a Neanderthal in relation to all other Neanderthals. But this time, instead of trying to obtain Thorin's chronology, our Neanderthal is placed in the model as a reference point limiting the potential age of all the other Neanderthals. We ask the computers to determine what happens if our Neanderthal wasn't 105 millennia old, but less than 50,000 years old? How does the model react? Will our sample be rejected as an aberration, an impossibility? If so, we'll have to continue our analyses to find out what's not working.

Setting up the model and then interpreting it would take several more months, but these analyses, placing Thorin as a reference point, would lead to the redefinition of the age of a group of other Neanderthals. This repositioning of Neanderthals from Spain, Central Europe and Eastern Europe provided particularly interesting results. On several of these sites, molecular chronologies had until then offered results that were sometimes clearly divergent from the dating of the archaeological levels from which they came. And these divergences, these inconsistencies, were systematically blamed on the archaeologists ... Poorly excavated sites, poorly understood contexts, poorly dated sites. But the insertion of Thorin into the statistical model as a reference point would modify and recalibrate the molecular chronologies of several Neanderthal sites, making it possible to bring molecular ages and archaeological dating into line. On these sites, the chronologies resulting from genetics now matched the ages of the archaeological levels in which they had been found. Not only did the model indicate to us that Thorin could perfectly well be less than 50,000 years old, but we had proof that the molecular clocks had until then been partially erroneous.

What was the source of these recurring errors resulting from these analytical models? The question still appears difficult to untangle but could well be correlated with differences in size between these popula-tions or with factors of the natural selection of individuals falsifying a simplistic statistical model that was perhaps too rigid to account for the history and real organization of these distant populations. Here, it is perhaps the particularities of the ancient Neanderthal ethnic and social structures that could directly impact statistical modelling.

What can we learn from this fight through time that oscillates, to within 80,000 years, between the 35th and 105th millennia? We could note that we are faced with a fairly rare situation in which a long process of demonstrations allows a human science to reposition apparently secure knowledge acquired through the so-called hard sciences.

But these approaches, these results, these fumblings, these questionings, are more than just the hard sciences foundering in the face of the human sciences: they are an integral part of what science is. As such, this episode reminds us that the scientific approach is unitary. In reality, there are neither hard sciences nor human sciences. Science is a certain way of conceiving and questioning the structure of reality. This way of constructing a view and analysing the materiality of the world around us sometimes requires high-performance techniques, heavy machinery, powerful mathematical models. But these powerful machines, exploiting the best of our technological capacities, can be calibrated, repositioned and profoundly challenged simply by using our senses, our thoughts, our words – with the help of certain logical structures of thought and, sometimes, with the help of a simple pair of tweezers bought at the local shop. As such, this long episode of research by the best international teams remarkably illustrates the unitary nature of science, not as a discipline, but as a way of thinking about the world. It also brings us face to face with teams who have never hesitated to challenge their own results, their approaches, their understandings, in the very long process of the production of knowledge.

We met Thorin in 2015 in the most astonishing way, on the ground, in the scrubland, as if time had forgotten to pass by in this corner of the forest; and it was only seven years later, in 2022, that we could begin to perceive his time, his ways and his traditions.

A last Neanderthal, in his place? Touching the unquantifiable

We can now find the words to describe Thorin – to say what he was and what he was not, what he experienced and what he tells us about his people, the last Neanderthals. But the anecdote of his own story, his life and his death, everything that constructs an individual in his reality, remains almost indefinable to us. A thousand little details could well, after seven more years of investigation, lead me to rewrite this chapter.

But not everything has been erased and fragments of his story have survived and have come down to us. In my defence, I have to say that my police investigation began only several tens of thousands of years after the events. But let's question a few observations. The remains of this body lay at the entrance to the cave, in the open air – not in the midst of what appear to be spaces for living, under the protection of the stone vault. Thorin was left in the open air, outside. But in these words, already, there's a subjectivity, a weighing up in the semantics of the terms used. When I say that he is outside, I imply that an outside could be distinguished from an inside among these populations and in this place. I could just as well say that he was left in the open air or under the stars. The simple observation that he was under or beyond the vault of the cave actually required a considerable amount of work, analysis and restitution in order to be able to model the morphology of this small cave as it was tens of millennia ago. The demonstration of this single small detail is in itself, already, a real feat of geological reading in this little cave. However, we can rely on this observation. Thorin lies well beyond what was, more than 42,000 years ago, the vault of this cave. But how can we define what this space represented in the imagination of this human group? We find here an incredible bric-a-brac of flints and bones. Real beds of objects abandoned by these Neanderthals on the ground. A rubbish bin covering the body? That would be very interesting. But we find the same indescribable bric-a-brac of flint muddled up in the cave, around roughly arranged hearths and surrounded by a few scattered stones. These abandoned objects, thousands of objects, may well raise questions. We know that the last moments of Neanderthal settlement in the cave represent the accumulation of several dozen seasonal occupations. These displays of flint are not the mark of a brief settlement but of a few generations of human lives. They had been coming here since the time of their grandfathers, or great-grandfathers. And perhaps well before. But they probably only spent a few days, a few weeks here, every year. It's both very long and very episodic. And this whole jumble of cut flints is a sign of these generations of anecdotes. They're everywhere, without any obvious arrangement that would allow us to easily recognize well-defined living spaces. I can see where they made the fire and the few objects lying near these two hearths but how was the daily life of the group organized here? There they are, women, men and children, since I find a few baby

73

teeth. The presence of baby teeth perhaps indicates a form of recurrence of these human passages since we don't lose a baby tooth every day. But the children are taking shape. I'd like to have seen these Neanderthal children play. Is it they who brought back this series of small quartz pebbles, these small round shapes of just a few centimetres diameter, some of which seem to me to be deliberately cut or broken? It's sometimes been noted that prehistoric woman was invisible in our writings and in our thoughts. But where are the children? These teeth betray the presence of the child but how can we see childhood? A child moves about, fidgets, occupies all the space. But forty-two millennia later we don't talk about it. Probably because here everything is uncertain. Rendered invisible in the mass of the uncertain it's no longer the subject of the archaeologist, nor that of science. And indeed we have nothing to go on – just a fraction of very small objects, tools of all shapes, so tiny, abandoned by the Neanderthals. We don't know if these miniature objects were the mark of these small hands. Nothing is less certain, neither these small pebbles, nor these few fossils, nor these small blocks of uncuttable (but very charming) flint. And our puzzling over these children shows how partial our readings are here. In thirty-two years at the Mandrin cave, I believe I have only been confronted with a single anecdotal item that I know to be the mark of childhood. And probably of a process of learning. It was a block of flint prepared by an artisan with great expertise, in order to be able to extract a single flint point. What we call a nucleus or core. The preparation was expert down to its smallest details, and showed remarkable workmanship. Neanderthals were sometimes experts in their gestures. And this one was indisputably a master carver. After this long preparation, the point could then be extracted in a single movement, a precisely positioned blow that marked the end of this artisanal process. This last gesture is the icing on the cake. It required a certain mastery. Oh, nothing extraordinary compared to the know-how of the carver. But the operation of extracting the stone point failed. And it was not so much the failure that was notable here – it happens to the best of us – as the fact that the one who failed persisted. And persisted. And persisted, each of his failed gestures breaking the rock a little more. Each clumsy blow was recorded forever in the material. Fifteen times he tried, fifteen times he failed. In actual fact, in this very fine flint, the first failure signalled the end of the game. The rock had been noted, had failed, and would provide nothing more. And these fifteen successive gestures, haphazard, a little clumsy where they struck, meant that whoever was trying to extract this point did not understand a certain number of very basic rules necessary for the exploitation of these flints. Clearly, the expert who had prepared the block at length was not the clumsy one who applied the last gesture and multiplied it in an irrational manner, each gesture making the block ever more fragile, more unexploitable. They were master and the apprentice. Parent and child? An anecdote across the millennia that spoke to us of those who know, of those who transmit and of the child who tries to understand, to reproduce. But this was the only time I saw both the child and the processes of transmission between generations. And wasn't some other scenario possible here? The master carver who wanted to extract his point and who, missing his shot, was overwhelmed by anger and hysteria and killed his block, voluntarily, in a fit of wrath? Or the child who, knowing where the carver kept his finely prepared blocks, took advantage of his absence to try his luck? Or a thousand little variations that remind us that our readings of these past moments are fragile. And what archaeology is looking for is never these anecdotes, which are rarely mentioned. Perhaps they're not there. Or perhaps the archaeologist doesn't see them – they're just a few more little flaws at the base of my block, after all. Should I read and interpret these few flaws? Is that my role? Defining an act, a tiny moment, instead of understanding a society? Should we work on major trends and reject everything else as the unapproachable element of what was being devised here? It's also rather as if these anecdotes had no right to exist, were no longer within the purview of science. Too uncertain, too unquantifiable. Too human. And here we face the heart of the matter. For the anecdote, which speaks to us of these children, speaks to us only of these little moments of nothing at all, does not in the least tell us about societies, their structure, their history. We have just entered an *elsewhere*, which transgresses what our views seem to be able to agree on. In our minds, all of this is no longer really science. But we will have to get down to it, since death is just one more anecdote. The last anecdote. Strong and powerful but an anecdote of nothing at all. A tiny moment when every-thing changes and the group will have to shake up, rethink all of its daily priorities. If we decide that the anecdote doesn't fall within the heart of our disciplines, we make ourselves incapable of being able to confront the definition of the tiniest gestures, those which, here, perhaps, are the only ones to make sense.

So, well then – are we confronted with a burial? No one knows. And yet, he was really lying there. The analysis only shows certain very limited movements of these bones through time. But is its position in this small natural pit, along this rock, a matter of chance? Is the body simply lying where this man found death? Or was he interred here by his people? In this hollow along the rock under the stars?

From flesh to bone

A few clues lead me to believe that these remains were found where Thorin was laid to rest. I just said 'laid to rest', taking a meaningful interpretative step. I don't think we are faced with remains lying on the ground. If that were the case, how could these remains not have been scattered by carnivores attracted by the smell of decomposing flesh? Those in his group probably couldn't stay here for more than a few weeks and had to continue their cycle of seasonal nomadism across their territory. The remains weren't scattered, and the elements of the head are found not far from the phalanges of the left hand. I don't yet know the precise original position of his body, and perhaps several hypotheses will have to be constructed, with varying degrees of confidence, as to the position of the dead man – on his back, on his side, lying in which direction? I don't know how to define this yet, but I know that if the elements weren't scattered randomly, at the whim of scavengers' jaws, it's because they were, in one way or another, protected. But here's the thing, the teeth of the skull aren't lying with the teeth of the mandible. They're separated by a few dozen centimetres. Just the right distance and probably in the right position for a skull rolling over on itself, leaving its jaw behind. But that's just what's surprising. The obvious protection of a body – something that comes instinctively to us in relation to our own traditions – would be to lay it in the ground, to bury it and to cover it. But these are *our* traditions and there are a thousand ways to behave in the face of death, from eating the whole or certain parts of the body and bones, to having it eaten by birds, or burning it, or hanging it, hiding it in corners, cracks, or trees, mummifying it, or bringing it back to its kinsmen from time to time by drying it and taking it out, all dressed up as if for a big night's celebration. Or cooking it all at once, or in several goes, or retrieving the head, or a bone, later, so as to always have it close to you. Four hundred millennia ago, in Atapuerca near Burgos in Spain, distant ancestors of the Neanderthals piled up the bodies of their dead at the bottom of a cave. Here, twenty-nine individuals were found. But were they deposited here through time, or did this group die during some sinister episode?

Interpretations are often fragile. As for Thorin, not only is interment in no way representative of humans facing death but if the goal was to protect the body from carnivores, the approach seems doomed to failure, as the hyena will come to dig it up. We must reverse our reasoning. At first glance, what is relevant here in our ways of thinking about these distant gestures is not simply that the group wanted to protect this body but that we are able to see that these remains could not, in this singular context, have reached us unless they had been protected.

It's this first step, this first observation, that will direct the first questions of our investigation. We're not inferring a deliberate act here, we're not interpreting a gesture, which is in any case indefinite: we are noting elements that induce a certain state of affairs – the history around this body. It must be hypothesized that something (an act? a singular process?) protected this body from carnivores, in the first moments after its death and until the disappearance of its flesh, the decomposition of which never fails to attract big carnivores in these landscapes where large fauna are so widely represented. It is therefore the question of the history of the body after Thorin's death, up until the disappearance of its perishable flesh. A body abandoned in the sun can decompose completely in about ten days. Buried, residues of skin, tendons or muscles can be preserved for several years. But, if Thorin died in winter, the polar climate in Europe during this period could have preserved the flesh for a very long time, even when exposed to the open air. Buried in frozen ground, such flesh can persist for centuries, if not millennia. So there is an uncertain time here that can range from a few days … to a few millennia. And it is over the entirety of this uncertain time that a body must be placed out of reach of carnivores in order to reach us, even partially, without being spread out over dozens of metres. The assessment, however imprecise, of a minimum time can give us interesting leads in the interpretation of the history of this body.

Nowadays, the melting of the great glaciers is releasing across Europe the archaeological remains trapped in its crevasses for millennia. These are no longer episodic discoveries: a whole arsenal of wood, leather and textiles is now being expelled by the accelerated melting of the ice. From Canada to Norway, via Italy and Austria, glaciers are closely monitored by the archaeologists who are gradually extracting their precious relics. Organic materials that have disappeared elsewhere are revealed and appear to us free of all the marks of time, as if they had just been abandoned. But they speak to us of almost all past ages, from nineteenth-century explorers to the Vikings and many millennia before – of all those who crossed mountains and passes, like the remarkable body of Ötzi. His body, mummified by the ice, was discovered by chance at an altitude of over three thousand metres in the Italian Alps. He had been lying for more than five millennia in the Similaun glacier, with a bow and quiver, his bearskin cap on his head and his deerskin shoes on his feet … In the immensities of Siberia, the remains of hunters from ancient prehistory emerge here and there, remains of those distant peoples of the mammoth that I mentioned in *The Naked Neanderthal*, on the margins of the world. Obviously, if any Siberian can immediately recognize the antiquity of the remains of a mammoth or a woolly rhinoceros when faced with a human body, the distinction between remains from the nineteenth century CE and the nineteenth millennium BCE is almost impossible to draw. On 19 September 1991, when Helmut and Erika Simon discovered Ötzi's body, they could only imagine it to be the remains of a mountaineer who had disappeared a few seasons earlier. It was the *gendarmes* who intervened first, and the films of the discovery show the attempts to extract this mummy from the ice encasing it using … ice axes! We are rarely prepared to confront the extraordinary. There is every reason to believe that the bodies of much older hunters emerge at a regular rate from the ground, petrified by the cold, and that they now lie – piously buried by their discoverers – under spruce crosses in peaceful Siberian villages.

Time often has little hold on the oldest remains. But the decomposition of a body can't be modelled so easily. It's never a mere question of time. We know that Thorin lived in a cold, very cold environment, more similar to the climate of the far north of Europe with reindeer and lemmings than to the scrubland that has spread across these Mediterranean areas for a dozen millennia. So we can't measure how long Thorin's transformation from flesh to bone took but this decay of the flesh was probably more gradual and took longer than it would be today at the entrance to this small cave, even if his death took place in fine weather when the ground was superficially thawed and bacterial activity was greater. Analysis of the remains of the small fauna – rodents, amphibians, reptiles – found in the archaeological levels of the Mandrin cave and very sensitive to climatic fluctuations, allows us to assess that around this forty-fifth millennium the frost periods spread over an average of 120 to 140 days per year. This places this Mediterranean area in a harsher atmosphere than that currently recorded at North Cape, well beyond the polar circle, on the extreme northern tip of the European continent at the top of Norway. Our Mediterranean area would also be comparable in many ways to the current temperatures of some of the regions of Siberia. The ground at the entrance to the cave could have been partially glaciated and digging a pit to place the body in could have been relatively complicated. Moreover, no digging has been observed in the area where these bones were found. However, the remains of this body appear in a basin along the rock; but analysis suggests that this small basin is natural and was already there before Thorin's death. The dimensions of this small natural pit were large enough to contain a body. Is this association between a body and a hole random or does it indicate the actions of the living after Thorin's death? The mark of these last Neanderthals in the face of death? It's almost impossible to decide without falling into a projection, an interpretation that would be no more robust than the sense that this association was random. Both propositions are equally fragile and it is largely on these fragilities that almost all Neanderthal burials have been widely discussed.

The mind in matter and the gaze versus reality

What could the location of a body in a pit or a depression tell us if we weren't able to recognize precise, thoughtful, irrational gestures there? Irrational because if humans were made of rational matter they would see the bodies of their dead merely as an inert mass to be cast away so as to remove the putrid odours, unless they took into consideration their energy value as a mass of cheap proteins that simply needed to be ingested. I have already explored the richness of all forms of cannibalism:

they are never limited to a question of mere proteins but always represent the mark of collective values of great complexity. And one can discern, underlying such acts, a confrontation of feelings that could well connect loving and eating, love and ingestion. It's like the laughter of cute little babies, so adorable you just want to gobble them up. And this relationship to the ingestion of the other could well cross over all cultural values and be part, not of the structures of this or that society but, more deeply, of our most universal, most natural behaviour, of our ethology, in an amazing ingestion of the other, a physical, symbolic, sentimental, moral and amorous ingestion. The total ingestion of the loved one could be an inescapable fusion in which one would like the matter of the other to drown completely in oneself. It's like a definition of love.

Everywhere we find the irrational, or the non-rational, and never humans thinking of the world in a statistical summing of its resources and potentialities. Evaluation is always subject to our values, our views, our thoughts, our subjectivities and, more deeply, our behaviours, our instincts, which we channel as best we can through our cultures. In life as in death and even in our cannibalistic forms. Everything is irrational, invested with unquantifiable values. And what if it were this irrationality that really and truly defined the human matter? Humans would not be a matter of tools, or bipedalism, or thought, or altruism but of our capacity to create a world that has no obvious echo in natural laws. Our capacity to reverse the reality of the world, to overturn the laws of nature in order to subject them, totally, to our gaze. Irrational, of course. 'The heart has its reasons that reason does not know' would not define the heart, but human matter, in its totality. In its depth. In its entirety. In reality, it's neither the heart nor God that Pascal is telling us about but our deepest nature. Our very structure as human beings, humans who have their reasons that reason does not know. And it's the way our thoughts turn things topsy-turvy, the way our creativity patterns all objective reality, that defines us totally, at each moment. But where, then, does this human nature come from? Where, in the millions of years that preceded us, could this trait of irrationality have appeared, this desire to bend reality to the sole force of our imaginations? Wouldn't all this simply be the original mark of what we are? Our true, our only original sin? In the creation of the very first of the first tools, more than three million years ago, can't we detect the irrational mark of a certain will – of a desire, a longing, to transform the world and turn it into a *something* else? A will to refuse reality and show that it is perhaps not so true, not so absolutely true, as all that. And perhaps it is enough, in this stroke of uncertain genius, to strike the rock with another rock, to transform it into something else that is sharper or different. Another reality, the emergence of the spirit in matter. The fruit of the confrontation of the gaze versus reality. And it is indeed the gaze that brought down the order of the world and forced nature into something else. A new state of things.

From the very first tool to the very first burial, it wasn't a case of stages of human evolution slowly leading towards what we are. There were just variations on the same reality that was there not only in embryo but already developing, from the first shock of these two stones commanding matter to bend to the mind. Already, in this gesture, all the irrationalities that forge human matter. All that, all that we are, would merely be the variation of the same profound impulse – something transformative and performative. Matter bending to the will, to the mind.

Mental processes on the edge of the hole

From the tool to the grave, an infinity of variations of the same irration-alities, transforming, bending, at will, any form of objective reality. Bending the real order of the world, the order that from all eternity obeys only the immutable laws of nature. Should we, as is often suggested, consider burial as a fundamental evolutionary step, perhaps more important than all the other irrational variations of human nature? More important perhaps, because this act, this moment, demonstrated the awareness of the existence of others, of their irreplaceability, making humans different from any other expression of the living world, from all those creatures that remained stuck in the natural realm? Something that, indisputably, would make us human beings?

And behind all these thoughts, these desires to differentiate ourselves, always this need to express that we are different from all the rest. Would this be the ego of humans seeking to stand out from the living, to see all other reality as inferior to our own nature? To distinguish ourselves from it as if we were, in the most obvious, the most indisputable way, superior matter? We would be this 'something else', this entity so different from everything that is alive and that hovers above the ether. Could this desire to conceive of ourselves in this way, to differentiate ourselves sharply from other living things, this obvious ego, represent one of our original weaknesses? Defining humans as the creatures who conceive that each being is irreplaceable – this could be a fine definition of humanity: flattering. Too beautiful? Too out of step with its realities, perhaps? A definition that provides us with the best image of what we are, differentiating us from the animal world no longer by objects or by techniques, as if this definition by the transformation of matter still positioned us too close to the reign of other living things and still kept us too close to the reign of creation. As if there were something vile and still too natural in associating the emergence of human beings with the creation of the first tool. As if the first burial extracted us, from above, from the realm of the living. Made us the astonishing creature that knows no echo, no parallel, no comparison with other creatures. There is in us, in our thoughts, even in our sciences, this powerful, transcendent will to detach ourselves from the natural realm – a will whose cultural value Descola has deciphered (for other societies position humans and non-humans differently). And yet, at the very heart of our Western conceptions, this singular etymological coincidence that sees these two terms constructed in *humus*, the Latin word for earth, as if to bury (*inhumer*) were not only to put something into the ground, but into man (*homme*). To make man, there would then be the germ, in the West too, of a kind of schizophrenia in which man is totally distinguished from nature and yet fully blends into it. As if, in the West too, the profound distinction between nature and culture were hesitant constructions, uncertain waltzes in which, perhaps, we cannot really position ourselves firmly.

And it is this meaning that we construct, on this notion of the first of interments. The first interment would therefore confer on us the nobility, the superiority that relegates all other living things to the rank of simple animated matter. As if we accepted that we were Nature but on the sole condition that we represented the only earthly consciousness. I think, I fear, that behind our words, what we need to look for is really this surplus of ego in all of us, this desire to distinguish ourselves deeply, totally. Could it be possible that in these ways of conceiving our long history we are trying to camouflage ourselves as divine creatures? Burial would then mark both a singular moment in the evolution of our societies and would demonstrate that other humanities were comparable to us – thus inviting those other creatures who also take care of their dead to our banquet of the 'more than nature'.

But we need to go further. Advance a few more steps. Scratch away a bit more at the surface of appearances. Are we really sure that it's only for these reasons and nothing else that burial is so important to understand the history of our humanities? Does the first of the burials really represent a singular moment in the history of human societies, as if the irrational gestures around death finally marked our extraction from the natural environment? We could then recognize a before and an after, as a stage from which those distant creatures, these prehumans and those other humans, were no longer mere simple bipeds, still a little limited at the level of the encephalon but had irrevocably tipped over into humanity and finally cleansed themselves of the natural? But could this not rather be, very crudely, because we recognize in this act of burial our own ways of being in the world? Not the ways of human beings but those of the societies in which we define ourselves as human beings? Much more than a remarkable evolutionary step, would there not be here too, yet again, that surplus of ego, eternally thirsty for differentiation, forever insatiable? The discovery of the tomb, perhaps, might not speak to us so much of those distant societies as of our need to talk about ourselves, to differentiate ourselves, to shout from the rooftops that we are much better, much more than what we were before. Yet again, we do not speak of our history, nor of the evolution of the distant societies that preceded us. We speak of ourselves. We are prisoners of our gaze. Of our need to distinguish ourselves, like children wanting to show their parents how grown up they are now. Could we be, finally, these eternal children, so unsure of ourselves that we have to show how our entire distant history demonstrates that we're different from all other living things, divergent from the entirety of the natural kingdom? How our natural matter has slowly sublimated into superior matter?

When, in 1999, we published in the journal *Science* the demonstration that some Neanderthals had been cannibals, some researchers welcomed the discovery but they criticized us, often tacitly, for having provided such an image of Neanderthals. For having associated them with savage, disgusting, bestial acts, undermining the slow work of rehabilitation of those populations. Such reactions actually showed us the opacity of our conceptions of what it means to be human, our

83

inability to confront ourselves deeply with any form of otherness. *The Naked Neanderthal* highlighted the mental processes in which we have tried to force Neanderthals into another Ourselves. And these reactions of rejection, disgust and reproach in the face of our demonstrations did not come from everyone but from the academic world, revealing the profound work that still needs to be done if our superficial views are to be left behind. The Epilogue to *The Naked Neanderthal*, 'Liberating the Creature', has few illusions about our ability to rethink the notion of humanity as a plural reality finally freed from conceptions closely linked to intellectual constructions that are culturally endorsed and totally construct us. It's likely that the task lies completely beyond us. The creature is our prisoner, just as we find ourselves, in turn, locked in our own mental cages. At this point we reach deeply chilling conclusions about human nature. Are we locked in cages without bars and that nevertheless guide, and even constrain, our view of the reality of the world? This is reminiscent of Aldous Huxley's essays. Building fictional scenarios and exploring the possibilities of our futures, the author of *Brave New World* sheds a harsh, penetrating light on certain structures of human nature. So is it fiction alone that dares to fully explore the deep mysteries of human matter and what unconsciously organizes the voluntary servitudes of our societies and our systems of government?

> They will do it by bypassing the sort of rational side of man and appealing to his subconscious and his deeper emotions, and his physiology even, and so, making him actually love his slavery. I mean, I think, this is the danger that actually people may be, in some ways, happy under the new regime, but that they will be happy in situations where they oughtn't to be happy.[4]

Huxley here questions some fundamental traits of human nature. Going beyond the observations on our nature, our ethological side, he explores the dangerous ways in which it interlocks with and subjects itself to our own cultural systems. These paths run through this tendency in our societies and in our very nature to accept a standardization in which all the submissions, not only voluntary but unconscious, of human societies seem to be constructed. The idea of these interlocking patterns of subservience brings together the thought of Huxley, who projects himself into possible futures, and of Michel Foucault, who explores an archaeology of the structures of domination: the two thinkers complement each other remarkably. Wouldn't the questioning of human matter lead us down a deeply ambiguous path – a path allowing us to lay bare the tools of domination that can govern human societies? Huxley and Foucault raise questions, point out risks and fragilities, and reveal our unconscious imprisonments. These mental servitudes impact everyone and the distortion of our perceptions does not affect the average citizen alone. Scientists, researchers, those who question the complexity of our matter, those who are supposed to shake the test tubes that govern our world are also stuck in this state of submission to our mental servitudes. We should shake the test tubes hard – but these experiments are uncertain and, above all, painful, every time. Questioning our abilities to understand our world by questioning and deconstructing each of the principles that we perceive as universal is an unpleasant experience. Dissociating ourselves from all shared thought, from all dominant thought, cannot lead to any form of recognition of the social body. It can only lead to isolation, removal and ostracism. And this step aside, this crossing, this transgression does not allow any backtracking. It means we engage in detours, deviations in which no destination is ever indicated. Instinctively, it is our nature always to refuse it.

But then, faced with these impossibilities of any profound freedom, in all these interlockings, these imprisonments of our human natures, what is humanity? And where is it hiding? If the first trace of humanity is neither the tool nor the burial, nor any object, nor any act, nor any of the forms of our material being, what then is it?

Might not the spark of humanity be precisely those tiny moments when thought finds itself freed from our conceptions of the world and offers us a glimpse of transcendence? Then, only freedom would be transgressive. Or rather, only the transgression of our ways, our conceptions, of our knowledge, of all our unconscious limits, would be freedom – creativity would be an act of humanity. Our freedom to conceive, outside our mental cages, would transform us, in a tiny instant, a luminous spark, into human matter.

There would be no humanity but a series of sparks. Rare moments of freedom, puffs of air, breaths where, for a tiny instant, but an instant of eternity, we would make humanity.

Finally, we would not be human. We would make humanity ... sometimes.

Here we are at the edge of the hole. Thinking about death. Thinking about burial. Thinking about humanity. It's not about measuring the depth of a hole, or demonstrating that this being, tens of thousands of years ago, was interred by his own people. It's first and foremost thinking about all of that.

Then comes the paramount question: why is this act of burial so consequential in our thoughts? Why does it seem to count so much, in the ancient history of human beings, if it is ultimately only one of the infinite variations of the irrationalities of human nature?

Perhaps, quite simply, because something in this act seems to represent how we define ourselves. In this case, aren't we blinded because we see in it the images that we have constructed to define our own reality? To define, in a somewhat narrow way, our values? And those values are perhaps not so absolute. They are not so universal. We aren't seeking the first burial, to respect the being we have lost forever. We are systematically seeking the first interment. The first interment, as if it were self-evident to inter one's dead. We seek acts that resemble in every way those of the societies most familiar to us. Interment, laying in the earth, is not in any way a universal act. Burned, eaten, given to animals or birds, hung in the air, left to dry in nets, thrown into the water, or invited to all the festive meals: the dead never simply leave. They are cumbersome. They occupy our heads and our dwellings. They spread out under our houses and in our streets. They're always there, in a thousand ways, whether we have to honour them or distrust them, mourn them or speak to them. Everything here is uncertain. The dead surpass their bodies, invade our thoughts, our societies, our neighbourhoods, our homes. They are, like us, matter and spirit. Like us, they are intrusive. And if we seek the nature of the living in the care of the dead, if we limit it to burial, to interment and if we try to detect the archaeological proof of a form of humanity, which is our way of being in the world, we are trying to discern in these distant populations what would make them *humanity*. But not humanity in the absolute but, in a more narrow way, what makes humanity in us, in our own view, in our societies. And this is just a mark of our own reflection.

Once again the creature, as well as all the other creatures of the distant past, escapes us completely. Tracking the origin of humanity, we are constantly tracking the only origin of what resembles *us*. Is that really reasonable? I don't know. But it's understandable, certainly.

For a few centimetres too many

Let's return to the edge of the hole, no longer that of our mental abysses but the one where Thorin was found. For now, the elements of this body essentially concern parts of the skull and the left hand. Almost all of the teeth are attested and arranged in two spaces: the upper teeth, from the upper jaw, are grouped together while the lower teeth, from the mandible, appear about thirty centimetres away. This distance between the upper and lower teeth marks a movement, a displacement that occurred after the decomposition of the flesh which held these rows of teeth together. We have seen that the time of the total decomposition of the flesh is uncertain, but it could be quite long in the cold environments of the last Ice Age. We then have two observations to make. The elements of this body were able to reach us because it was not destroyed, scattered, consumed by the many carnivores that populated these regions at the time. But after the flesh was freed, certain movements can be recognized. These movements show the distance between the mandible and the skull, which rolls away. If it rolls, it's because it *can* roll. This sentence isn't here to make you smile kindly as at a truism: rather, if this body, placed in a natural grave, had been interred there, covered with earth, then the skull and the mandible embedded in the sediments would not have been able to move like this. And this distance of thirty centimetres corresponds quite precisely to the distance that this skull would have travelled by rolling over on itself, once freed from its flesh. A remarkable paradox here appears; the body must have been protected since the elements that constitute it were preserved from carnivores but the body was probably not buried. If it had been buried, this movement of the rolling skull could not have taken place. The fact that the skull can go gallivanting around is, however, a classic observation when opening old tombs, when the body is laid in empty spaces, in a coffin, a vault, or a dolmen where the head is sometimes found between the legs of the deceased. Cross-referencing of the clues around the position of these remains demonstrates that the body could well have been protected from large carnivores without having been buried. Located along a rocky ledge, in a small natural basin, we have to imagine that, trapped in its depression, in the small natural setting in which we found it, the body was covered with perishable materials that protected it from the action of carnivores. Arrangements of trunks could well have protected the body here, preserving it in an environment that was both closed and open, positioned in an empty space allowing the skull to roll over after the decomposition of the flesh. These clues would lead us to the conclusion that the preservation and subsequent movement of this body resulted from the actions of the members of the dead man's group. If it wasn't an interment, since it wasn't actually laid in the earth, it was indeed a burial. A set of acts of care shown by the living towards the dead. Remarkable, rare gestures, beautiful anecdotes that speak of a fraction of the traditions, thoughts and immaterial conceptions of that distant Neanderthal group. And we remember that Thorin had a very singular object a few centimetres from his skull and his left hand. Not only is it the most beautiful flint blade found in this archaeological level, but it also reveals a disturbing observation. At first glance, this object does not in the least resemble any of the tools left to us by these Neanderthals. When I first picked it up, I thought of one of the Proto-Aurignacian blades. It was long believed that this culture marked the first colonization of Europe by *Sapiens* somewhere around the forty-second millennium. But it actually found expression a good twelve millennia after the first major migratory phases of the middle of the fiftieth millennium. The first major wave of settlement was in reality simply the last. On the other hand, it is this last wave that coincided with the evaporation of the Neanderthal populations, who disappeared like snow in the sunlight. A strange conjunction of two singular events that we still do not know how to connect together.

Thorin appeared on the ground, at the first stroke of our brushes. He lay precisely in the top level of our archaeological records. Here we find the rare remains of the modern humans from forty-two thousand years ago, those who may have encountered the last Neanderthals. The flint blade was found a few hours before Thorin's first teeth emerged. I picked it up to analyse its technological information. A flint can be read like a book. It records a large part of its history. The artisan's gestures are written in it. We can recognize the knowledge of the carver and the traditions of his population. These gestures do not speak solely of its creator's skill. They also tell us about the culture of his entire group.

There are ways of being, there is knowledge and there is a style. And if at first glance I thought I was dealing with one of these *Sapiens* blades from the forty-second millennium, the second glance immediately reoriented my gaze. This object had nothing of *Sapiens* about it except its style. The style without the technique. The technological traditions discernible in this blade were clearly Neanderthal. During this phase of settlement of the European continent, Neanderthals and modern humans sometimes produced objects that could resemble each other. What we have here are transitional industries, as if the Neanderthals had first been swept away culturally before being swept away biologically. Transitional industries would then be like a little Neanderthal leap before the big plunge, the last cultural stop before extinction. Surprising, isn't it? Three hundred millennia of technological stasis followed by a single jump before disappearance, like one final realization just before they crashed into the depths of their evolutionary impasse … I use the term 'evolutionary impasse' here as if Neanderthals were the actors of their disappearance. But the remarkable organization of these societies, their adaptation to all environments, doesn't suggest any impasse. And in the Rhône Valley none of the great transitional industries of Western Europe have ever been documented. Here, Neanderthal died elegantly, remaining faithful to what he had been, for all eternity. No process of destructuring or of cultural reorganization is documented in this region. Neanderthals are there and then they're no longer there. End of story. Move along, there's nothing to see here. No slow evolutionary transformation. No slow cultural agony before the disappearance of the bodies. No brilliant last leap before extinction. On the other hand, here we have a body. And a blade. At first glance, the blade has the taste and smell of *Sapiens* traditions. But, at second glance, the object is clearly Neanderthal. Beautiful work but Neanderthal work. And Neanderthal doesn't need to cross paths with *Sapiens* to learn how to make blades. He's been producing them for tens of millennia, from time to time, when he feels like it. And it's within this 'when he feels like it' that the great divergence between these two humanities probably begins. The Neanderthals were playing with their knowledge of materials, following the order of the materials and *Sapiens* were applying their recursive formulas, quoting them endlessly like an obligatory recitation. And under the gaze that I posited in *The Naked Neanderthal,* we realize that the distinctions, the divisions, the particularities of these two humanities do not emerge in any way from the technologies of these two societies but from their ways of registering their presence in the world. Of understanding the world. It's not the presence of a blade that bothers me, moreover. It's actually its style. This blade really has a *Sapiens* smell. A bit as if its artisan had been able to have in his hand, or not far from his eyes, a *Sapiens* blade that he had reproduced in his own way, keeping just the style, creating a fairly convincing overall vision but reinterpreting the object through the filter of his own technological traditions, or rather through the filter of his own ways of being in the world. The remarkable object, abandoned just a few centimetres from Thorin's left hand, could well correspond to a real mortuary deposit. But what a deposit! An object from between two worlds, belonging in its techniques to Neanderthal traditions and dependent in its style on *Sapiens* traditions. It must also be recognized that you have to be remarkably gifted in the mastery of your crafts to perfectly reproduce a style by following technological procedures, since sequences of gestures have no connection with the copied style. There is in these arrangements a form of raw genius. A bit like a painter who cheats with the reality of the world and produces illusions of materials, textures, volumes and spaces simply by applying a few lines. A few colours. In this sense, our Neanderthal is a remarkable impressionist who very skilfully plays with the illusions of the world. Yet here we only have a stone blade. But it's this object from between two worlds that we find associated with the body of one of the last representatives of this population, before extinction. Here is an object heavy with meaning and which might draw a direct line between these two humanities.

But this whole construction that directly links two humanities must probably be abandoned. If what I've just described is both accurate and factual, the few centimetres separating the flint blade from the hand represent an unbridgeable chasm. Was this blade really associated with the dead man? 'There's many a slip 'twixt the cup and the lip', as my grandmother Yvette gleefully repeated every time I presented her with some obvious fact. Admittedly, she'd experienced the Occupation, its deprivation, the lack of everything. Endowed with an iron character, my old granny knew the emptiness of all promises. The history she had lived through made her gaze swivel systematically into the world of uncertainty, into a universe where nothing is ever certain, neither tomorrow, nor yesterday. Positioned near this hand, the blade wasn't *in* the hand. Oh, maybe it had been there. And even if it wasn't held in the curl of the fingers, it had perhaps been placed near this body.

Maybe. Maybe not.

And those few centimetres can never be crossed. We have a hand and a blade of stone. Which rub shoulders, as it were, without perhaps ever having met. Those few centimetres are so important. This uncertainty is the whole story of Neanderthals and *Sapiens*. Across Europe their remains haunt the ground, side by side, without us ever being able to know if they ever crossed hands. We have only the certainty of extinction and of replacement a few centimetres away. But those few centimetres are impassable. Together they lie between zero and infinity.

And if this blade is remarkable in itself, its modern style is only an uncertain detail. Does it mark an encounter? Perhaps, a bit like those aboriginal representations that represent sailboats between two paintings from the dreamtime. The sailboat drifts along between the kangaroos and the magic symbols driven into the rock. Neither the kangaroo nor the sailboat, which wander on the same wall, have ever sailed together. Facing the wall we are confronted with an art of appositions. These images were simply the material support of tales that were told. Without the words that dressed them, that built them, they're merely skeletons whitened by light. This stone blade does not escape the art of appositions. It will no longer tell us anything if the aborigine does not explain its profound nature to us. But in its way of being, between two worlds, it could well trace the existence of visual contact. The Neanderthal world gazing into the *Sapiens* world or not.

Magic against the intangibility of time.

And here, we are faced with a confusing, disturbing observation. Even when we find the body of one of the last Neanderthals, and even when this body is surrounded by objects from between two worlds, we still can say nothing. We can affirm nothing. We can conclude nothing. But we have opened doors. We have explored possibilities. We considered uncertain moments. This body is indeed here. These strange movements and arrangements in space are here, too. But what readings can we give them? And how can we evaluate their scope and meaning?

All our logics can be reversed and it is ourselves that we must be wary of. Ourselves we must distrust. Fear that our reading is only *our* reading. As if the remarkable event of this man's death did not really exist. As if only our readings, our views remained. And in writing this, I'm well aware that it's due to these errors that thought has been rejected. That it's in the face of the rich ambiguity in the reading of an event that thought has been discredited. As if thought were too uncertain, too shifting, too dangerous to be used. It is in these observations that people decided to no longer invoke the verb 'think', replacing it with lines of calculations, statistics, measurements and radiometry, or alignments of genes and proteins. But all these palliatives to thought are not thought itself. And they don't remedy anything at all. If we refuse to think because thought is dangerous, no model, no calculation, will replace it. Thought will simply be dead. And yet, the reassuring calculations, the edifices of the hard sciences, could well be as fragile, as false, as uncertain, as the thoughts they hope to replace because of the uncertainty of all thought. We have seen Thorin wandering from the thirty-fourth to the one hundred and fifth millennium, depending on the model, without these oceans of uncertainties suggesting that we reject our statistical tools and our mathematical models. Thought is no less uncertain than these models. All these calculations, all these models, are simply the toolboxes allowing us to construct our thoughts. The exercise cannot therefore stop at the construction of remarkable mathematical models. On the contrary, it's here that it must begin to be exercised. It must borrow from all these disciplines so that it can finally free itself from them. Thought is both the guide and the next step. The moment when we manage to mix the alignments of numbers and the raw observations as they appear in the grease of the caves. This is a very uncertain equation. But it's the only one that works. The only one that *can* work. I believe so … I believe it, because this experience, this confrontation with fieldwork, is in no way interchangeable. All the calculations, all the measurements, all the fieldwork archives analysed a posteriori no longer fully testify to what archaeologists have experienced and what they are trying to testify to. We would like archaeology to be a simple, technological, reversible action, reinterpretable after the fact. But archaeology is none of that. Or it is so only imperfectly. We take many photos, many notes, we exchange many glances, but once the mandible is extracted from the matrix that has contained it for tens of millennia there is no going back.

This confrontation with the body of one of the last Neanderthals fascinated me. And this fascination did not stop me hating the actions I had to accomplish. I was clearing and destroying. I was extracting and forever removing artifacts from their in situ locations and placing them into a box. I was archiving, scanning, modelling and labelling. Trying to transform the physical reality of these remains into a virtual, modelled reality. Infinite care, extreme attention and very slow actions nevertheless result in the destruction of the original arrangements of these human remains as witnesses.

The stone tools that we find in archaeological levels frozen for tens of millennia function like time capsules. The tens of millennia separating us from their artisans have no substance in reality. This cut flint that I extract from the ground is deposited just where its artisan abandoned it. The remains of this fire still mark the colour and texture of its soil. Faced with amazingly well-preserved archaeological levels time becomes totally impalpable. We see the objects, we see their arrangements taking shape before our eyes, marking each of the gestures of these hunters. The event could have taken place forty or one hundred millennia ago, time here is totally intangible. Time becomes a construction. An illusion, but one without texture, because sometimes, in these archaeological actions, nothing connects us to time anymore. There's no thickness. Nothing is distinct. Everything appears to us in the most instantaneous way. As if the object had been directly handed to us by its Neanderthal artisan. But here too, or perhaps above all, our minds are totally playing with us. It does happen that time can be totally inconsistent, transparent but we can't see *through* time. Its thickness has simply become invisible to our eyes. The mind does the rest. In reality, time is most often an invisible thickness and gives us the remarkable illusion of its non-existence. It's like when walking on a large rock slab you suddenly discover the traces of a thousand small fossilized shells. You're walking on a beach forty-five million years old. The shells are still coloured. It looks as if they'd just been pushed along by the waves and even the ripples of the sand are still visible, right there, under our very eyes. This invisible time, this illusion of past dunes, or the impression of being able to take this stone tool as if it had been handed to you personally by a Neanderthal, form the common ground of all those who work on the past. But magic is not this intangibility of time. The magic is in the researcher's ability to draw the thickness of a time that is nevertheless invisible.

So here we are, facing this body. We know that it was probably placed in a basin after its death. That this basin was probably covered with perishable materials, wood, forming the vault in which certain parts of the body would subsist and travel through the millennia. We also know that objects could have been deposited there. Objects that could well evoke the world of others. And with much more certainty we know, finally, that none of these sentences is an affirmation. These sentences, my words, are always hypotheses, readings. They are possibilities.

No affirmation holds true. Neanderthals systematically impel us to abjure our certainties, reminding us of this sentence attributed to Galileo: '*E pur si muove*' ('And yet it – the earth – turns.') Don't take this last sentence as an affirmation that my doubts shouldn't be taken too seriously but rather as an additional layer of doubt. And, faced with these piles of questions, I really and sincerely know that I do not know. And we play through these words, reconstructing remarkable anecdotes, like the gestures of this child who was seen cutting flint and getting it wrong fifteen times. The archaeologist is poorly equipped, not really *cut out* to read all these old anecdotes properly. From the discoveries of childhood to the death of humans, all these little moments of life, all these little instants facing death, escape us. A little as if these moments, so precious, forging the deepest raw material of all humanity, had never counted.

I can't see the detail of a single death and yet I am tempted to speak of the death of all other humans as if it was a warning. Like it was a moment of confusion. We don't see well through time. Maybe we don't see well at all, even in the single present moment.

The first of the last contacts?

So, did the first of the last contacts ever take place? What do the crafts of the last Neanderthals evoke for us; all those transitional industries, those strange cultures that dance between Neanderthal knowledge and modern traditions, speaking to us of contacts, exchanges, acculturations, of the traces left by one humanity on another? Perhaps these traces, perhaps all these traces, are only illusions. Perhaps these distant clues are just *our* illusions. Our fantasies about the encounter between human beings and the creature. Perhaps all our hypotheses about these acculturated, culturally modified Neanderthals affected by their contact with *Sapiens* are false – are our myths. What if the first of the last contacts were not there? What if Neanderthals had been neither acculturated, nor assimilated, nor transformed by their uncertain, unquantifiable, indescribable encounter with *Sapiens*? What if our understanding of the last Neanderthal crafts were completely distorted? What if Neanderthals were not where we expected them to be? What if all these Neanderthal craft traditions showing the acculturation of these populations were simply not the work of Neanderthals?

We still have to scrape away our thoughts, our imaginations, to remove the fog from the windows and look more closely. Perhaps Neanderthals can still surprise us, perhaps they disappeared without ever having changed anything they did for all eternity. Perhaps there was no acculturation. No Neanderthal invention of any form of our outdated modernities, nothing but this good old Mousterian, the good old Neanderthal crafts composed of shards and, among a thousand others, of those dear old Levallois technologies whose origins, as we have seen, can be traced back more than a million years to somewhere on the slopes of a volcano lost in the deserts of Anatolia. Neanderthals who were always identical, incorruptible, impervious to external influences that perhaps never happened. Perhaps the last Neanderthal isn't the one we thought, or the way we imagined. But all this is invisible – invisible because we haven't dared to think outside of paths that have been traced too long.

Like those invisible boats that crossed the seas to Australia more than fifty millennia ago. Those shadows of boats that we cannot see. They've disappeared, completely disappeared, decomposed, melted into invisible dust. But it was easy here, in our imaginations, to look at the invisible. By the simple, unwavering observation – since the aboriginals can still look you straight in the eye – that these humans must have crossed, in one way or another, those vast expanses of sea. Here we have the first invisible encounter of humanity with a new continent; there, the encounter of two distant humanities on a given territory.

Always, the invisible.

But unlike Australia, in our old fiftieth or fortieth millennium Europe, nothing is easy to observe. Nothing is easy to imagine. The equation is far too complicated and includes an infinite number of unknown variables.

The genes are the easiest, almost mathematical part to analyse. And we can always tinker with mathematics. It's not like human beings. Human beings are hard to classify. They're not all neat and clean. Not very logical. Really, they're not presentable. They're messy and even a bit dirty. They smell. While genes and mathematics are practical. They're clear. They're understandable. Mathematics is clean. It looks good. It gets published in the major international scientific journals. But we don't do science with technology and we don't understand humanity with statistical concepts that are too neatly ordered. It's not reasonable. Human matter is based on the irrational. Humans smell bad, they ooze, they never do what they should, they don't correspond to anything really quantifiable. They don't fit so easily into a box. I'm not sure it's reasonable to study something irrational and smelly with mathematics. Faced with this unpresentable, ugly material, wouldn't the desire to take refuge in genetic, statistical, radio-digital tools, all very clean, very presentable, very quantified, fitting neatly into boxes, already represent a form of prurience, a kind of refusal or negation of what we are? Because the reality of human flesh, of its psyche, well, how can I put it ...? Yes, it's uncertain. It's filthy and inconsistent.

And yet, the history of societies engaged in these processes of extinction and replacement remains mathematically indefinable to us, thus abandoning us to a rather blissful expectancy. We can't as yet understand these events, or even document them, much less define the technological and social structures of these societies. Contact, first contact, is generally limited to highlighting, across Europe, craft traditions that are thought, or believed, to be Neanderthal, which evolve and invent, perhaps through contact with *Sapiens*, new ways of being in the world. New ways, new cultures, new names. Neanderthal traditions following the cultural fashions of the new world are called Uluzzian, Châtelperronian, Bohunician, Jerzmanovician, and Lincombian and are recognized by many other names, sometimes even more exotic, which have in common nothing but the fact that they are generally attributed to the last Neanderthals, Neanderthals acculturated in the *Sapiens* fashion, and that they end in *-ian* ... When we speak of human societies, *-ian* most often sounds like a mark of the uncertain, the indefinite, like the Pre-Columbian, which does not speak of any very specific cultural group but of entire sections of populations in the Americas before Columbus, before the process of European colonization. The Neanderthals, also saddled with this -*ian*, form part of these uncertain zones by their very name. But rest assured, all the other prehistoric cultures also rhyme in -*ian*, and doubt is elegantly shared by all these old, uncertain societies.

That would be the contact: all those traditions in -*ian* that emerge across Europe, a sudden flowering of unexpected technological innovations after hundreds of millennia of Neanderthal technological continuities. But perhaps all these -*ian*s tell us nothing about the last Neanderthal moments, tell us nothing about extinction, and tell us *absolutely* nothing about the last Neanderthal. And perhaps none of these flints in the cultural fashion of the new world were ever cut by Neanderthals. Neither by the last Neanderthals, nor by those before.

Perhaps we need to rethink everything.

Three waves breaking against the intellectual Titanic

Thorin was truly one of the last Neanderthals. And here, these populations died out without change. Without movement. I don't mean that they died without a cry, without a death rattle. I mean that they remained until their final moments what they had been for all eternity; they died without transforming themselves. Not that they didn't see the Others but these Others didn't transform them. There's this little blade, which perhaps doesn't tell us anything but perhaps tells us that they did see and recognized a difference. But these societies in the Rhône Valley remain very Neanderthal. Structurally Neanderthal in their ways of being in the world. Totally Neanderthal until their final moments. No acculturation. No transformation. None of those tipping points that would see the Neanderthals become like the Other. You remember, all these names in -*ian*. No Châtelperronian here. No Uluzzian. No Bohunician, and none of the other thingummies in -*ian* that describe the moment when Neanderthals apparently changed just before disappearing – apparently invented, or copied, or mutated into *Sapiens* technologies. Here, in the Rhône Valley, they turn the page without warning. Here, they're associated with their old Neanderthal technologies until the end. Not even a comma between the old and the new world. I exist and then I no longer exist. I do not transit to anything. I am and I fade away just as I was for all eternity.

As if this death of humanity represented merely a snap of the fingers. A snapshot.

And, when I think about it, I wonder if what I see in the Rhône Valley isn't the tree that conceals the forest. If in reality no Neanderthal society ever transited or transformed. I wonder if all these societies in *-ian* are in reality not the mark of Neanderthals in any way, but all the work of *Sapiens*. The question has been much debated since we have so few human remains.

In the absence of remains it was the hard sciences that settled the question and they inclined rather towards Neanderthals, it seems. With the help of DNA extraction, ancient proteins and extensive dating, the old sites of the Châtelperronian fell into Neanderthal hands, profoundly impacting both the image we can have of these populations and the processes by which this humanity became extinct. But these analyses were carried out on objects extracted from caves decades ago. These are old excavations. Very old excavations. And remember, we can never go back. It's an astonishing paradox but archaeology doesn't lend itself very well to going back in time. Whatever the quality of the recordings, what was missed, what was not perceived in the fieldwork, cannot be reinterpreted. Should not be reinterpreted. These attributions to Neanderthals involve painful, delicate transformations. Yet it is on these a posteriori constructions that some have attributed the Châtelperronian to Neanderthals and others to *Sapiens*.

Faced with these irreconcilable paradoxes, my own gaze has not been directly drawn to these Châtelperronian ensembles. It's just too uncertain. Too subject to the gaze, to analyses, to transformations, to illiteracy – since here the power of the word has been swept away, replaced by measurements and quantifications. I must admit that faced with these mechanical rumblings, I fled. My thoughts don't walk in step and are wary of rows of numbers that are too neatly aligned. So I fixed my gaze from the outside. Instead of proposing a rereading of this or that Châtelperronian site, I tried to understand these productions as a metasystem, tracking the ramifications of this phenomenon across time and space. My work on Harvard University in 2016 allowed me to examine flint crafts identical to those of my Neronian in the Rhône Valley. I thus recognized a remarkable framework that united the two opposite shores of the Mediterranean over more than three thousand kilometres.

Here, in the Mediterranean Levant, I recognized my Rhône Valley crafts of the fifty-fourth millennium. This community of technological traditions right across Mediterranean spaces could not be random. And here, on the slopes of Mount Lebanon, I knew that *Sapiens* was the author of these crafts. It was on these comparative bases that I initially proposed the existence of these astonishing trans-Mediterranean lines of descent, positing the hypothesis that *Sapiens* must have reached the shores of Western Europe much earlier than had previously been postulated. So it was a wager. A predictive wager. Without safety nets. My hypothesis pushed back the arrival of modern humans in the west of the continent by twelve millennia. Quite a leap in time. The analysis of human teeth found in the Mandrin cave would later confirm the hypothesis established on the basis of the comparative analysis of these crafts at both ends of the Mediterranean.

But I didn't want to focus on this single moment of the Neronian, on this true 'people of the point', as their technologies were entirely oriented towards obtaining fine standardized flint points. I had to understand, contextualize the meaning of this astonishing craft of flint points. But the investigation into the origin of these technologies would run into a dead end. I opened the drawers containing the flints belonging to the older populations in the hope of being able to trace the origin of these amazing technologies. But the ancient levels showed very different traditions, as if these Neronian points suddenly appeared in the archaeological records without any local antecedent, even in the eastern Mediterranean. The populations preceding the 'peoples of the point' were the heirs of other traditions. The origin of the Neronian was probably to be sought elsewhere. Perhaps in other regions of the Mediterranean East. But perhaps much further, towards the high plateaus and expanses of Central Asia. All the great migrations come from this immense reservoir of Central Asia, from the peoples of the Bronze Age to the Huns and the Mongols advancing towards the West. It's always the immense Asian expanses that we see washed up on the banks of the West of the continent. While in our conceptions of the great migratory waves Africa is generally singled out, it's actually the vast Asian expanses that, in the history of old Europe, bathed and fed the main phases of population of our continent. And in these vast Asian expanses, *Sapiens* has been loafing about for a very long time. We have been able to distinguish traces of it here and there for perhaps two hundred millennia. From the shores of the eastern Mediterranean to southeast Asia, *Sapiens* left us the marks of their passage, of their settlements, without it ever being possible to define very precisely, region by region, the moments when they definitively occupied these vast expanses. In the Eurasian west, the Neronian probably corresponds to one of these innumerable Asian waves crashing into the west of the world. Not that *Sapiens* isn't ultimately an African creature – it *is*, biologically, in its distant origins, in its flesh. But it could well be that the populations that finally turned towards Europe were profoundly Asian. In the middle of the fiftieth millennium their traditions, their cultures, their technologies could well point to greater Asia. Maybe we are not here faced with a simple African migration flowing directly towards the west of Eurasia. In any case, if the collections from the slopes of Mount Lebanon allowed me to pinpoint the incredible similarity of the distant Lebanese and Rhône societies, they didn't allow me to define the origin of my Neronian. It will probably be necessary to look elsewhere to try to untangle some of this immense historical skein.

On the other hand, the analysis of flints from more recent levels, later than the Neronian as it can be discerned in the Mediterranean East, would open me up to unexpected horizons. By looking at the flints, archaeological level after archaeological level, we can see the traditions of these human societies evolve, very slowly, very gradually, through time. If the origin of the first technologically modern societies could not be recognized on the Lebanese deposit of Ksar Akil, their future is wonderfully documented there. By analysing the archaeological levels of the oldest to the most recent human occupations, we can see how these societies gradually evolve, change and modify their traditions. It's an amazing experience as we then have the strange sensation of being confronted with a biological entity evolving before our eyes, mutation after mutation. Amazing, because if we now know how biological evolutions and mutations work, the evolution of techniques, societies and traditions remains much more mysterious. There's no natural law, no Darwinism applying to objects manufactured by humans. The laws of nature and the laws of techniques do not obey the same properties. By definition, nothing about techniques falls under the laws of nature. On this point, we have nature and culture. We see, we understand how selection and the realities of the natural environment will influence the future of species. But when we are faced with a tradition, a culture, knowledge, understanding, prohibitions, no one knows what can guide the mutations of human societies and how populations always end up switching from one tradition to another. The countless theories about the collapse of civilizations only tell us about particular cases. And if they tell us about what changes, what collapses, they do not tell us about what remains. They can sometimes try to discern why things change but they do not really tell us about how things change. About what changes and what persists. And when we approach it, we simply have to note it, rather as if no law, no rule, could be established, could be modelled. We cannot discern the cultural lineages that nevertheless connect human societies. We cannot define any laws structuring the future of these evolutionary lineages. Or if we engage in this astonishing experiment, we know that the exception will, systematically, have every reason to be the rule.

If no rule can be established, if no law seems to be able to define the evolutions of human societies, the Mount Lebanon collections illustrated some remarkable processes, very gradual changes in their technological traditions – very slow mutations, archaeological level after archaeological level, like a biological evolution, or rather like a plant that grows but gradually becomes something else. Reading these flints was fascinating. I didn't just recognize the old traditions of points from the Rhône Valley. These archaeological excavations had also revealed, several metres above the Neronian, settlements in every way comparable to the European Proto-Aurignacian. The 'Proto' is the traditions of the forty-second millennium, that great wave of settlement that can be recognized almost everywhere on the European continent, seeming to sweep away all Neanderthal societies. So I had before my eyes, expressed on a good fifteen archaeological levels, all the slow, gradual evolutions leading from the traditions of the Neronian to the traditions of the Proto-Aurignacian. I could see and define how societies entirely focused on the production of flint points would become societies focused on the production of large pointed bladelets. It was as if these points were gradually evolving, like biological entities, becoming even more pointed, finer bladelets. And these gradual evolutions therefore suggested that these two traditions, the Neronian and the Proto-Aurignacian, were one and the same. The same tradition separated by twelve millennia of slow evolution. I could see and define how the traditions of the fifty-fourth millennium *Sapiens* would slowly transform into something else. Here, on the slopes of Mount Lebanon, there was no break from the Neronian to the Proto-Aurignacian. But in the Rhône Valley I was not faced with a gradual evolution of one cultural entity into another cultural entity. In the Rhône Valley I only saw the end of the processes. The moment when everything was over. Just waves of settlements finally coming to the west of the world in two very distinct waves of *Sapiens*. In the Rhône Valley I didn't see any society transforming from the Neronian to the Proto-Aurignacian but two phases of cave occupation separated by twelve millennia. Here, on the banks of the Rhône, not only were these societies separated by a deep temporal thickness, but I had no possibility of finding any continuity between the Neronian and the Proto-Aurignacian. And between these two modern pulsations I could recognize five major phases of Neanderthal settlements intercalating and marking the return of those fascinating creatures to the Rhône territory. *Sapiens* at the top. *Sapiens* at the bottom. Neanderthals everywhere else. No possible local evolution. No break in continuity, here, between the Neronian and the Proto. But over there, on the eastern banks of the Mediterranean, it was a different story. Not only could I see how, from the point of view of these crafts, societies slowly shifted from one tradition to another but this passage from the Neronian to the Proto-Aurignacian was not linear. If I could characterize the slow transformation of the technologies of these populations, these processes of cultural mutation did not document a simple shift from the Neronian to the Proto-Aurignacian. I could discern an intermediate stage. An unexpected stage. There was a phase intercalated. The Neronian first became something else before slowly transforming into the Proto-Aurignacian. And in the most unexpected way this something else had the taste and smell of the Châtelperronian. I had before my eyes the progressive mutations leading from the Neronian to the Proto-Aurignacian via the Châtelperronian. You remember, the Châtelperronian, those technological traditions of the southwest of France and Burgundy that also vaguely linger in the north of the Iberian peninsula. Those traditions that – we were promised – are Neanderthal, the familiar between-two-worlds that saw Neanderthals switch over to beautiful modernity in *Sapiens* fashion before disappearing. And in Ksar Akil, these technological traditions that have the taste and smell of the Châtelperronian were obviously the work of our dear *Sapiens*. A double burial was found there. Two young kids. Two little *Sapiens* who died sometime between the forty-second and fifty-fourth millennia. But on the slopes of Mount Lebanon all those modern industries, from the Neronian to the Proto, are obviously signed *Sapiens*.

Now that's astonishing. No. Now that's *really* astonishing.

So, in Mandrin, I had two periods of the same tradition. The beginning and the end, the Neronian, and the Proto. Two distinct waves of *Sapiens* populations in Europe. Two waves expressed twelve millennia apart but two waves coming from a single cultural substrate. And this Levantine cultural substrate was clearly also the one that would give birth to the Châtelperronian. In Europe, three distinct cultural traditions expressing themselves in different geographical areas. In the Levant, one and the same *Sapiens* tradition evolving through time. What we had, distributed in space, dispersed horizontally and discontinuously across the geographies of Western Europe, was what we had recorded, vertically and continuously through time in the eastern Mediterranean. And if you've followed me closely, you'll realize that I reject the attribution of the Châtelperronian to Neanderthals and that I'm telling you about three great waves of *Sapiens* settlement moving towards the European continent. Three great waves expressing themselves over twelve millennia.

If my proposition is even slightly accurate then all the historical schemata that we have constructed to account for the settlement of Europe by *Sapiens* are simply false. Completely false. Everything needs to be rewritten. Everything needs to be rethought from top to bottom.

There was no late *Sapiens* migration in the forty-second millennium but a plurality of waves of settlements spanning millennia across European regions. No sudden disappearance of Neanderthals upon contact with the arrival of *Sapiens* in Europe. The Protoaurignacian, long considered the first major wave of *Sapiens* settlement on the continent, was in reality only the last. Everything took place over a much longer time than we thought. An infinitely more dilated time – and yet without any obvious impact on Neanderthal cultural traditions. It could well be that this key period did not actually see any of the technological and cultural changes previously recognized within Neanderthal societies. And this is perhaps the most important point. The Châtelperronian would then be only a variation of the Levantine *Sapiens* traditions. But then, what about the great transition and all those cultures in -*ian* marking the profound transformation of Neanderthal societies before their extinction?

The hypothesis I have just presented to you suggests that what I saw in the Rhône Valley, this sudden extinction of Neanderthal societies without any mutation of their technological traditions, without any transformation of their societies, could well, in reality, be not an exception but represent the rule of an incredible historical process: the extinction of a certain humanity.

The Rhône exception would be just the visible part of an immense interpretative iceberg. And the vast corpus of hypotheses built up over the last fifty years on the Neanderthal extinction in Europe would be nothing more than an incredible intellectual Titanic. And what if, as incredible as it may seem, we had missed the essential part of this immense story of the replacement of humanity?

The hypothesis that I am here posing develops in a rather particular context that has seen, over the last ten years, the first modern industries attributed to Neanderthals gradually fall into the hands of *Sapiens*. This happened first with the Uluzzian, a sort of counterpart of the Châtelperronian attested in Italy and Greece, then with the Bachokirian in Bulgaria and, finally, in 2022, with the Neronian. In the meantime, the similarity of the old Bohunician of Central Europe with Levantine traditions attributed to *Sapiens* had already meant that its attribution to modern humans was seriously considered. So now was it the turn of the Châtelperronian? The technological connections that can be sensed with the Levantine space most probably represent a very serious warning. The Châtelperronian, one of the most important transitional industries in Western Europe, would – due to the technological structure of its crafts – have very little chance of being included in the field of Neanderthal traditions of Western Europe. The Châtelperronian would fall entirely into the hands of the *Sapiens*. We thought we could recognize in the west of the continent two major waves of *Sapiens* settlement, separated by twelve long millennia during which Neanderthal societies once again reigned supreme over the west of the continent. But these data suggest that there were actually three major waves of settlement. We would therefore have an almost continuous presence of *Sapiens* populations in Western Europe from the fifty-fourth millennium until the extinction of the Neanderthals, twelve millennia later. Three waves of *Sapiens* settlement and at the same time no cultural mutation of Neanderthal societies. The hypothesis of such continuity of settlements of the continent is astonishing. It is unexpected and it profoundly rewrites the historical structure of this shift in humanity. But I think that it is, structurally, the most probable hypothesis. The most astonishing but the most probable. This hypothesis is based on the analysis of the deep logic of the crafts we can observe and on a structural, comparative and trans-Mediterranean approach. We don't as yet hold the human remains or their genetic traces that would allow us to close the propositions that I have just established. So it's a reading, a hypothesis, that I'm proposing here. And it's a prediction. The characteristic of science is to be able to pose hypotheses that can be tested, confirmed or refuted, sometimes with a time lag of decades.

I therefore predict that this understanding, this reading of flint crafts, is better able to diagnose the signature of human societies than, at time T, the disciplines resulting from the so-called hard sciences and pointing to Neanderthals as the creators of the Châtelperronian. I am betting on *Sapiens* for this culture and wagering that the human sciences, structural analyses and trans-Mediterranean comparative approaches offer greater analytical power for the recognition of human societies than a disembodied use of radiometric measurements, genetics or isotopes.

But then, in this great domino game of toppling modern traditions outside the Neanderthal reign what is ultimately left for Neanderthals?

Well … nothing … Probably nothing.

I mean, none of all that.

This whole story of transitional industries, of the mutations of Neanderthal societies before extinction, could well be nothing more than a huge intellectual dead-end. A myth. A construction. And I can already see my colleagues getting together to reject such a hypothesis en masse. Attacking it from every possible angle. But I'm afraid we're going to have to swallow this pill. No point making a flap about it. I fear that the ship's going to sink even if the band plays on.

Now, history will decide.

But, if my hypothesis turns out to be correct, all our explanatory schemata for one of the greatest extinctions of humanity ever recorded are simply false. Entirely false.

There'd be nothing left for Neanderthal. But 'nothing', that's fine. I mean it's better. Because if Neanderthals never did any of this then it's a matter of the utmost urgency to be able to rid them of all these trappings. It's time to rid them of Us, of our constructions, of our fantasies, of our gazes. We could finally look at them as they really were.

So ... Maybe everything has to be rewritten. And it's much better that way.

We would finally have Neanderthal, naked.

Genetic surprises

This last chapter's shaking a bit but I hope there are still passengers on board happy to have seen the immense ship of our intellectual constructions set sail on a journey that could well lead it, not to sink – since the depths are also uncertain – but to head out towards unexplored horizons.

So maybe, finally, the Neanderthals died while remaining what they'd always been. True to themselves, they'd have been poetic enough to die while remaining what they were for all eternity. There was no transformation of Neanderthals into another version of Ourselves before extinction. The creatures would finally escape from us. This liberation is a unique opportunity to try to think differently, not only about that distant humanity, but also about our way of apprehending it. To reinvoke *the word* as the primary tool for thinking about all forms of humanity, to relegate all the disciplines that are not founded in the first instance on thought, on the word, on subtle analysis, to the rank of mere tools, a rank they should never have left. The remarkable tools of dating, protein analysis and genetics are just tools. They can only accompany thought. But in expressing this, I'm not establishing a hierarchy. I know that the best people in each of these disciplines have arrived at the same conclusions. If, at first, the dazzling advances in dating methods and even more so in palaeogenetics were primarily based on the skilled use of extraordinarily expensive machines, these disciplines have now reached a major turning point in their history. These tools are rapidly becoming more democratic and it is no longer the mastery of increasingly sophisticated machines that will make the difference in terms of scientific impact but the ability of these teams to integrate their way of questioning the raw materials with the issues slowly built up by archaeology. And it is now this humanized use of these remarkable tools that will probably enable us to ask the most innovative, the most promising, the most interesting questions. This is my conviction. And my hope.

And again, it was just where these remarkable tools encountered our knowledge about Neanderthal societies that something unexpected would emerge that morning in January 2022. The teams in Copenhagen continued to work away at Thorin's genetic data, trying to track down the reasons why their first models could have led to such errors regarding the real age of our Neanderthal. Placing Thorin in front of all the genetic data obtained on Neanderthals, the Danish teams found themselves confronted with an even more astonishing reality.

Dear Ludovic,

We've now made good progress in the analysis of Thorin's genome. Very surprisingly, our models reveal a remarkable genetic divergence of Thorin from all the Neanderthal populations recognized in Western Europe. His lineage seems to separate from the lineage of Neanderthals in Western Europe 105,000 years ago. This means that Thorin's population shows about sixty millennia of divergence from classic Neanderthal populations. This is an extremely big divergence and particularly surprising because during these 60,000 years of divergence the Thorin population doesn't show any genetic exchange with other Neanderthal populations, including those located a few hundred kilometres from the Mandrin cave. These data are surprising but we have carried out a vast corpus of analyses that allow us to affirm the absence of any genetic exchange since the one hundred and fifth millennium.

On the other hand, we discern a very close proximity between Thorin and the Neanderthal from Forbes' Quarry in the south of the Iberian Peninsula. These two individuals are incredibly close genetically and they are very clearly isolated from all other Neanderthal populations. As you know, Forbes' Quarry is attributed to an ancient population between seventy and one hundred and twenty years old, corresponding to the age that we ourselves had attributed to Thorin in the first analysis.

We're very curious in Copenhagen to receive your first impressions and your interpretation of these unexpected results.

Kind regards,

Martin

Sometimes there are letters we weren't expecting to receive. This message from the teams of geneticists in Copenhagen opened me up to unexplored scientific possibilities. Thorin, who had initially challenged our dating capabilities, now questioned our models of Neanderthal population distribution. We were faced with a population unknown to the genetic records recognized in Western Europe during this period. But what did these results mean? Did they suggest that Thorin did not in fact belong to a recent Neanderthal population but to ancient groups, earlier than the hundredth millennium? Was it necessary to rethink everything, reject all of our analyses again, and question our Neanderthal anew? And how were we to understand this very clear connection with the extreme tip of southern Europe?

The development of genetics in the last decade has made it possible to recognize different Neanderthal populations in which European Neanderthals are quite clearly distinguished from their eastern counterparts. In Europe, the most classic branch of Neanderthals can be recognized throughout the territories of the west of the continent, from Belgium via France to Spain. These populations seem to show a remarkable continuity of settlement from the hundred and fifth millennium until the Neanderthal extinction. For more than sixty millennia these Neanderthals seem to have been very homogeneous, very close genetically, suggesting a population that was probably small on the continental scale.

A population quite distinct from these Western European groups has been recognized in recent years. These eastern Neanderthals are recognized more than five thousand kilometres from Western Europe, on the southern side of Siberia, in the valleys of the Altai, those mountains on the borders of Russia, Mongolia and China. The complexity of Neanderthal populations has been recognized not only through genetic analyses established on bones but also through the extraction of genetic traces preserved directly in the soils of caves in Europe and Siberia. These astonishing traces of DNA recorded in the old sediments of caves suggest that the landscape of Neanderthal populations has changed considerably over time, perhaps under the effects of different migratory waves within these populations themselves. The deepest of these waves seems discernible some one hundred and five thousand years ago, giving rise to different Neanderthal branches occupying Western and Central Europe, the Caucasus and Siberia. However, the deep structure of these moments of population exchange and replacement escapes us almost entirely. These distinctions of populations, these movements, these exchanges were not expected. Either these movements were not discernible by archaeology or we never considered exploring the connections that could have existed so long ago in territories separated by such vast distances. And yet, archaeology had long distinguished the particularities of Neanderthal traditions in these different areas of Eurasia but without being able to distinguish or precisely recognize the surges of activity in which not only human or cultural exchanges occurred but also real population replacements. If these moments of mutation, these breaks, were not expected within the Neanderthal populations themselves, archaeology nevertheless proves capable of discerning these moments of divisions in human societies. The most remarkable and striking of these breaks in the history of human societies probably concerns the replacement of Neanderthal societies by *Sapiens* populations, even if what we perceive of them can only be a stylization of these distant events. Until 2022, we thought that this process of population replacement took place around the forty-second millennium, before realizing that this period did not mark the beginning but the end of a very long process that remained below archaeology's lines of visibility. This process had actually begun twelve millennia before. These twelve millennia represented the incredible thickness of an invisible time that we had not been able to document. Until then, and with a twelve-millennia gap, we had defined a moment of replacement, with a time before (Neanderthal), an in-between (with the social and cultural mutation of Neanderthal populations), and an after (now fully modern). A simple, linear intellectual construction, structured like a biological entity that changed from the simplest to the most complex, from the Archaic to the Modern. From 0 to 1, with a simple 'time between two worlds', a 0.5 in which Neanderthals expressed, in a techno-logical apotheosis, their own cultural reversal before extinction. In such diagrams, we can distinguish not historical processes in their genesis and their history, their interactions, but outcomes in which societies do not display a precise and well-defined history. We can distinguish the ends of journeys, the ends of evolutionary lines, and focus our attention just on the moments when in reality everything had already been played out – twelve millennia earlier.

But well before the great extinction, before the insoluble death of this humanity, in the interplay of displacements and replacements that involved the Neanderthal populations alone, and were confined to them, neither the analysis of the rich data from archaeology, nor even anthropology or genetics as yet allow us to understand the historical processes in which the balance between populations seems to tilt. The history and cultural anthropology of Neanderthal societies remain entirely to be explored. Probably these immense distances, the thousands of kilometres separating Western Europe from Siberia were perceived as powerful barriers that our scientific questions could not really cross. Barriers that Neanderthals could overstep but that science could not really explore. It was all too far away. Would those geographical, ecological and cultural environments be too distant to be precisely explored as far as the Neanderthals were concerned, even though in *Sapiens* the possibility of distant connections, crossing and connecting continents, is conceived as a possibility, something one might well expect?

Finally, it is our own political barriers that have long distinguished worlds of research and such investigations could not have been easily embarked upon while our societies were mired in the strategic confrontation of the great political blocs of the twentieth century. In this case, it was our barriers, no longer intellectual, but social, political and strategic, that imprisoned the east and west of Eurasia in poorly communicating spheres, creating distorting mirrors through which the Other became infinitely more distant than it really was. Here politics disunites, deliberately disunites but only because it is trapped in itself, no longer able to distinguish what nevertheless fully unites our own humanity in an inseparable mass. Our divergences are mere details in the face of the divergences, in their own lifetimes, of extinct humanities. The migrations I am talking about here were not our migrations, those which occupy – far too much – politics and the media. The migrations and replacements of the old plural humanities were in no way comparable to what we call migrations and replacements today. Now, we form, in our flesh, just one, and only one, humanity. And our cultural differences are like pleasant anecdotes in the face of the divergences of our past humanities. The replacements of humanity, the real replacements, did exist but they are unlike anything our societies can experience. These distant migrations, even if they were so inadequately investigated, were profound; here, they resulted not only in certain unexpected cultural mutations but also in real eradications, the precise content of which we have yet to define. How do human beings die? And indeed, well before expiring, how do they encounter one another?

Tens of millennia before the *Sapiens* migrations towards Europe, in the heart of the old Eurasian continent, the very directions of the movements of the different Neanderthal lineages are still not perceptible to us. Were Eastern Neanderthals moving towards Europe? Or were European Neanderthals moving towards southern Siberia? What happened in Spain after the hundred and fifth millennium to the Neanderthal populations replaced by Neanderthals who were clearly genetically differentiated? These questions remain largely open. However, on the current basis of genetics the populations in Western Europe after this hundred and fifth millennium seem to be unified. The Neanderthals show unity and great genetic homogeneity, suggesting that in this immense straight line before extinction, in their last sixty millennia, these populations seem to have finally stabilized in their different territories from Europe to Siberia. As if these Neanderthal movements were expressed during these distant phases before the last glaciations and the climatic collapses, the great outbursts of cold, had frozen the situation of these humanities. Did the Neanderthals then freeze, as if packed into geographical spaces rendered immutable? The proposition seems counter-intuitive, however. Temperate environments before the hundredth millennium see the development of unthinkable forest covers. These immense primary forests could hardly have facilitated the movement of human populations. However, it was these forest peoples who seem to have left their imprint on the great movements of their times. The immense grassy expanses would subsequently draw a new geography. The landscape geography of Eurasia during the great cold ages offered landscapes stretching out to infinity, that should have facilitated, or even invited, populations to move around. To explore. People do not move around in endless primary forests the same way as in a prairie unifying all of Eurasia and transforming it into expanses of short grass that the eye can easily gaze over in all directions. But if the climate directly modifies natural environments human beings can't be captured so easily by such explanations. We can make models, play on temperatures, modify and specify all the associated biotopes. But no analysis has ever been able to establish a convincing correlation demonstrating that environmental mutations structured the history of human societies. While many studies explore such paths in immense statistical models, a quick look at what constructs the very notion of humanity suggests that human beings have never complied with any of these models. Human beings are the unexpected, the irrational. And the intangible reasons why they decide to cross forests, rivers, oceans and mountains are not accessible using such simple algorithms. When humans decide to challenge space, crossing thousands of kilometres, none of these movements finds an echo in any population model. Human movements cannot be thought of mathematically. Human movements can only be observed. The models distance us, in their spirit, in their very nature, from all human beings. No model can explain any of the diffusions of planetary populations. Neither hot nor cold, neither hunger nor appetite. The colonization of Australia at the very moment when the Neronian *Sapiens* were exploring and settling Western Europe can only be posed as an observation. But why at this precise moment? Why this synchronicity at both ends of the immense expanses of Eurasia? We are faced with observations for which no model offers any relevance. Not that no explanation is accessible to us but the investigation should probably be limited to the comparison of empirical data. And to their subtle, qualitative processing, imposed and constructed through the mind and the word. Humans are restless and never predictable. They are unlikely creatures and they scorn distribution models. Here they are, colonizing polar spaces and sauntering across deserts, against all models, against all obvious rational logic. But the Inuit and the Tuareg live there. It's just a fact. Just as the flight of the bumblebee, heavy, round, massive, doesn't obey in any way the smooth laws of aerodynamics, human populations can't be modelled in their futures or in their actions. They are statistically uncontrollable. They're there or they're not. This isn't much use to the archaeologist but it allows us to think of humanities in certain ways. It allows us, as it were, to ask the questions we ask of these creatures through time.

These populations comply only with observable facts. The aborigines live where they do and their presence is a fact that opposes the logic of any model.

For Neanderthals, the fact seems to have been this: from Spain to Belgium and throughout Western Europe, Neanderthals display a single, very homogeneous population. And the genetic variations between Neanderthals from northern Europe and those from France or the Iberian Peninsula appear remarkably low. In this area of Western Europe, Neanderthal populations moving away from this European lineage are discernible only in ancient phases, prior to the hundredth millennium. The complexity of these ancient populations subsequently becomes invisible. After the hundred and fifth millennium European populations seem to have been fairly stable across the continent and were deployed in very different geographical and climatic contexts.

So what did the analyses of the Copenhagen geneticists mean? Did the coincidence of the Thorin lineage and Forbes' Quarry in the extreme southern tip of Spain imply that the Mandrin Neanderthal was in reality much older than we had deduced? Should we call into question our seven years of research and collaboration with the best international teams?

The genetic data on Gibraltar had been published in 2019 by the excellent teams at the Max Planck Institute, including Svante Pääbo, who won the Nobel Prize in Medicine three years later. This was a fine piece of work on collections that are difficult to analyse. DNA is not well preserved in warm climates. And this research focused on a skull that had been extracted decades ago from the cave in which it had lain for tens of millennia. The Iberian data were therefore particularly valuable and their successful extraction was quite remarkable. They were therefore telling us something important, which I had to take into account, with the utmost seriousness. The geneticists concluded that the genes in the Forbes' Quarry skull linked it to populations sixty to one hundred and twenty thousand years old, much more than to more recent Neanderthals. They also noted the very probable precedence of Forbes' Quarry over other Neanderthals discovered in Eastern Europe. This Iberian Neanderthal must have been older than the almost complete body of a two-year-old child discovered at Mezmaiskaya 1 in 1993 in the Russian mountains of the Caucasus. This child was dated as sixty to seventy thousand years old. The chronological range of the Forbes' Quarry skull could therefore be further narrowed and statistically situated somewhere between the seventieth and hundred and twentieth millennia. A very wide range of fifty millennia, certainly, but a range lying between twenty-five and seventy millennia before Thorin. We had here much more than a simple hop, skip and jump across the millennia. The very close proximity between Thorin and Forbes' Quarry was possible only on the condition that these two individuals, the Mandrin Neanderthal and the Gibraltar Neanderthal, came from a very close, if not identical, time period. However, this study proposed, in its words, an alternative solution: the possibility of the existence of a distinct population *within* the late European Neanderthals. But this possibility was perceived as highly unlikely improbability and was mentioned in passing, in an isolated sentence, almost on principle – as an improbable alternative, to put it mildly, to the conclusion of this study, since all the European data argue and start from the observation that the late Neanderthals in Europe represent, everywhere, only a single population. If this alternative interpretation seemed swept aside by this study, these data allowed two quite divergent readings as to the interpretation of this skull at the extreme southern tip of Europe.

Why and how could two such dissimilar scenarios be extracted from the genetic readings?

Forbes' Quarry left us a relatively complete, well-preserved skull. It was discovered in 1848 in the quarry where the rocks of the imposing Rock of Gibraltar were worked. This skull is therefore in reality one of the oldest Neanderthal remains ever discovered, extracted eight years before the remains from the Neander Valley. It is quite complete and its Neanderthal morphologies are relatively low-key. Its relatively soft bone reliefs did not allow the existence of a humanity distinct from ours to be recognized. A few years later, the remains of the Neander Valley would yield bones whose bone morphologies were clearly more pronounced. The recognition of a humanity hitherto unknown was finally established. Neanderthal man was finally discovered. Today, with our hindsight on these populations, the relatively soft character of the bone reliefs of Forbes' Quarry seems quite compatible with the antiquity of this individual. Neanderthalization is indeed an evolutionary, gradual process. The most recent individuals are therefore those presenting the most pronounced morphological traits within these populations. All in all, these data suggested that Forbes' Quarry had indeed every reason to be quite an ancient individual. And the suggestion of an age between seventy and one hundred and twenty thousand years old seemed pretty persuasive.

But one salient point emerged from my discussions with the geneticists in Copenhagen. Thorin and Forbes' Quarry are very close. Genetically.

And they also have every reason to be the same age. The chronology of one of these two Neanderthals is wrong and must therefore be abandoned.

Who's wrong? Place your bets...

What Gibraltar tells us above all is a story of archaeology. The lucky blow of a crowbar in a quarry in the northern face of the Rock of Gibraltar in March 1848. The context of those old Neanderthals is always that. The essential feature in all our bodies is a good old blow from a pickaxe or the explosion of a well-gauged dose of dynamite. But after one hundred and seventy-four years, Thorin's blow of the pickaxe was the object of millions of tweezer strokes, being analysed by the best teams in the world, calibrating and tilting the dates of the other Neanderthals so often discovered thanks to ancient booms of dynamite.

It wasn't Thorin who needed to be rethought. It was Gibraltar that needed to totter and tilt.

We need to understand. We mustn't just read the writings of others. We must go and see things in situ, facing, forty-five millennia after the peoples of Thorin, the geographies, the spaces, and the strange connections of humanity that have been suddenly revealed.

We must always see with our own eyes, let our minds gather and establish their connections that no statistics can build. Our minds and the confrontation with the data will lead us into the unexpected.

We must go to Gibraltar.

The stone ship

We must go to Gibraltar, but not by plane. We must confront the landscapes. The geographies. See all the spaces slowly unfold from the Pyrenees to the extreme southern tip of Europe. This path is an integral part of thought. It's one of the primary structures of all understanding. A few days' drive is enough, today, to travel from the Pyrenean foothills to Gibraltar. If the Pyrenees form a fortified line separating this vast peninsula from the rest of the European continent, the peninsula is largely made up of mountain ranges and high plateaus. Spain ranks second in Europe in terms of its average altitude, just behind Switzerland, and at an average altitude twice that of France. Its mountain fronts surround the peninsula and overlook the Mediterranean, from the Pyrenees to the southern tip of the country, where I am heading to meet the scientific teams investigating the Neanderthal cavities. These cavities record the southernmost occupations ever colonized by Neanderthals. And neither the mountains overlooking the Mediterranean nor the high plateaus allow us to recognize obvious geographical continuities with the rest of Europe. We are far from the vast Rhône Valley where Thorin lay. The Rhône represents a natural geographical area of circulation. A furrow linking the Mediterranean to continental spaces in the same way that, geographically, the Iberian Peninsula is isolated from the rest of the continent. In the early 1990s, the existence of a border distinguishing the first modern humans and the last Neanderthals in the heart of the peninsula was proposed for this reason. This hypothesis suggests that it was not the Pyrenees that perhaps divided the world of the last Neanderthals but, just to the south, the small Ebro Valley that flows from west to east across Spain. This southwards shift of the natural and historical borders between human societies is largely explained by the discovery of Châtelperronian and early Aurignacian occupations north of the Ebro River. And the first modern societies of the Aurignacian, known everywhere else across Europe, seem to have left no trace of their existence south of the Ebro. We could then document across this true subcontinent of the Iberian massif the persistence of Neanderthal populations long after their extinction from the European continent. But carbon-14 techniques are fragile and particularly sensitive to heat. Fossil bones are affected by the climate. Dating and genetics rarely allow the development of robust and indisputable measurements here. It is by the presence or absence of the first modern cultures of the Aurignacian and the Châtelperronian that such a model can be constructed. Did Neanderthal populations persist across the Iberian subcontinent? The suggestion, combining geography, history and human societies is fascinating and must be openly explored. Let's bear in mind, however, that the Ebro is a river only in name. Its flow rate is less than a quarter of that of the Rhône, at the same level as French rivers of fairly modest size such as the Meuse or the Saône. However, the modesty of such a river does not allow us to discuss its value as a border built by distant human societies. But this border was less physical than social, like the one that seems to have emerged in the Rhône Valley after the Neronian occupations, after the hundred and seventy-second millennium. Suddenly, after the departure of *Sapiens*, no one passed or crossed over. No population seems to have been able to exploit the western banks of the great river anymore. And here the powerful flow of the river is irrelevant. During these Ice Ages, was it not essentially frozen during the long winter season? The imaginary borders of human beings like to model themselves on the major geographical events of their landscapes, not because these borders do not obey geography in any way, but on the contrary because they emphasize in spatial terms the line of human wills. Massifs and rivers may well be natural, but the way they are interpreted and inserted into the mental universe of humans does not so much relate to geography as to the mythology of the living. One of the major traits of human populations in their relationship to geography, in their constructions of territories, relates essentially to socially invested values. The rooting of human territory in an immutable geography primarily traces the mental space of populations. These spaces are never limited to landscapes in which resources are exploited. Human territories correspond above all to the geographies in which the bodies of their dead have been abandoned, bodies from which one cannot so easily distance oneself. Human territories obey their roots in the lands of their deceased.

Facing the Spanish highlands, crossing those mountains, my thoughts process down the path to the southernmost tip of the European continent.

Positioned at the end of all lands, at the very end of our continent, Gibraltar is still a British overseas territory. In August 1704, the Rock fell to the British after three days of fighting between British forces united with Dutch troops. The fall of the Rock made it possible to seize this unique space, granting control of the passage between the Mediterranean and the Atlantic. Seventy-five years later, from 1779 to 1783, in the midst of the American War of Independence, the French and the Spanish united to lay siege to the Rock, in an attempt to overthrow British control. But, while French intervention would help to free America from the British, the immense rock presented itself as an imposing natural citadel and could not fall so easily. Gibraltar would remain definitively under British command. But the political situation of this major strategic place did not impose itself in any negotiation but in a balance of power favourable to Gibraltar. The Rock constitutes an impregnable fortress. Gibraltar is a mountain of stone rising abruptly from the sea. From the eighteenth to the twentieth century, the very hard limestone was gradually pierced with tunnels and underground fortresses. Sixty-five years after the French siege, in 1848, the northern front of the Rock was being worked in quarries under military command, in order to obtain the material to rebuild and reinforce the walls of the city. It was under the aptly-named Edmund Flint, captain of the Royal Navy, that an astonishing skull was unearthed. While there is probably no connection to be made between the discovery of this archaeological treasure by Captain Flint and the emergence thirty-five years later, in 1883, of another Captain Flint in Robert Louis Stevenson's *Treasure Island*, one could almost believe, following the path of this exceptional discovery, that it is as much myth as real history. The antiquity of the skull sent to London was immediately established. In the middle of the nineteenth century the discovery was of prime importance. Darwin would say a few years later in one of his letters that 'the Gibraltar skull is magnificent', but he would not mention this discovery in his treatise on *The Origin of Species* in 1859 and would touch on it only very briefly in 1871 in *The Descent of Man*. Note that it was at this pivotal moment, when neither Neanderthals, nor archaeology, nor the notion of the evolution of species were yet precisely theorized, that this skull emerged from quarrymen labouring under military command. This work was aimed at the strategic reinforcement of Gibraltar. How did this skull appear and how was it extracted from the Forbes' Quarry cave? The context of the discovery is not precisely documented and dynamite would not be invented by Alfred Nobel until eighteen years later, in 1866. The work of the quarrymen relied solely on the strength of their arms and on the mastery of metal tools, picks, chisels and crowbars.

It is while thinking of this very singular history and geography that I finally see the immense rock emerge. At the extreme southern tip of the peninsula, I am confronted with Gibraltar. The Rock suddenly emerges in the middle of the waves like an immense stone vessel. A remarkable outgrowth – not just a rock, but a peak, a cape, a whole peninsula. The breathtaking limestone ridge plunges into the sea in a straight line, six kilometres long and just a little over a kilometre wide. A great line rising straight up into the sky and plunging from north to south into the sea.

Overlooking the waves, the Rock rises abruptly to an altitude of 426 metres. Here history and geography meet myth. These immense cliffs in the water are Heracles and his columns, connecting and distinguishing the worlds of the far west.

It so happens that the Gibraltar National Museum is directed by Dr Clive Finlayson, an archaeologist who has been involved in the archaeological excavations of the Gibraltar caves for around thirty years. For the Rock of Gibraltar is also an immense receptacle that has left traces of some quite exceptional Neanderthal occupations. The discovery in 1848 of one of the most complete Neanderthal skulls is in reality only the tip of the iceberg. This immense stone vessel contains world-famous Neanderthal cavities, listed as World Heritage Sites.

I've never met Clive Finlayson, but I've been familiar with the work of his teams since his announcement in 2006 of the latest dating in the world of archaeological levels attributed to Neanderthals. Clive suggests that these populations persist on Gibraltar – some fourteen millennia after the disappearance of all Neanderthal traces across Eurasia. Or rather, throughout classical Eurasia, because my own research on the Arctic Circle would reveal five years later the perpetuation, in the same period, of Neanderthal traditions precisely when we turn to the other end of the continent, to the Russian Arctic Circle. But neither on Gibraltar nor on the Arctic Circle was the slightest Neanderthal bone vestige discovered. We are confronted with flint tools characteristic of Neanderthal crafts. And these technologies were abandoned by *Sapiens* well before the fiftieth millennium. In *The Naked Neanderthal*, I question the strange connection between polar spaces and the permanence of these ancient craft traditions. In the twenty-eighth millennium, in Gibraltar as in Byzovaya, are we faced, in the polar cold, with the continuation not only of old technological knowledge, but also with the ultimate survival of Neanderthal populations? No answer can be given here; it would be premature. In both cases, we are, quite surprisingly, facing the most peripheral geographical spaces of the European continent. The ends of the world, positioned precisely beyond the continental spaces, far from all the exchanges, all the circulations of old Eurasia. And the persistence of these artisanal practices, precisely at the two geographical margins of the continent, cannot fail to raise questions.

Gone with the Wind

I had an appointment in Gibraltar with Clive and teams from Arte who wished to record our meeting. The director wanted this meeting to be a real meeting. A non-fictional exchange that would not stage our interviews but record them in real time. Clive and I accepted this condition of a head-to-head and the possibility of recording the real world. So we'd reduced our discussions to a single interview with the producer and director of this Arte documentary. This conversation allowed me to tell Clive the reason for my scientific interest in Gibraltar, to express the unexpected connections revealed by genetics that united Gibraltar to the Rhône Valley in the last millennia of the Neanderthal reign. A union against all odds and to the exclusion of all other Neanderthals in Western Europe. I'd forwarded the details of our genetic analyses and our scientific conclusions to Clive. A few days later Clive sent me a laconic message.

> Dear Ludovic, the implications of these studies are simply wonderful. We must meet.
> Best wishes,
> Clive

So we played along with the rules of the game, conducting a meeting recorded by the cameras, but a real meeting in which we could directly discuss the meaning and implications of this unexpected discovery.

This morning, 27 October 2022, I have an appointment at the southern end of the Rock in the Punta de Europa car park. I find myself near the lighthouse at the tip of Europe, not on a cool late October morning, but in an exceptional maritime mildness. Géraldine Finlayson, head of the scientific laboratory at the Gibraltar National Museum and Clive's partner, has arranged to meet me and accompany me to the Gorham and Vanguard Caves, where their teams are currently excavating. We go to the military zone on the eastern side of the Rock, put on white helmets and begin the descent towards the maritime caves at the very bottom, directly licked by the sea. We're about a hundred metres above the caves, but there are still more than three hundred metres of cliff hanging over us. Clive will join us with the archaeological teams later in the morning, directly in the Vanguard Cave below. A few hundred steps further down, I find myself in front of a fascinating landscape. Huge stone mouths. Caves several dozen metres wide and high open their mouths wide into the waves of the Mediterranean. Following the sandy slope that climbs towards the bottom of the cave, we can clearly see staircase-shaped excavations. These steps reveal the excavations being carried out by archaeologists in the sand trapped in the cavern. A few dozen minutes later I hear voices on the stairs going down along the outer cliff. Clive arrives, accompanied by his excavators. We wave to each other from afar. We have a lot to discuss if we are to understand this astonishing puzzle, of which genetics assures us that we hold two perfectly fitting pieces.

We enter together into the vast natural cathedral of Vanguard Cave under the flights of thousands of swifts that nest in the crevices of these stone vaults. At the age of nearly seventy, Clive's life has been filled with Neanderthal explorations in the bowels of the Rock of Gibraltar. His announcement in 2006 regarding the very late survival of Neanderthals at the southern tip of Europe has largely fuelled scientific thinking and the vision that the general public now has of the question of the last Neanderthals. This announcement was an elegant extension of the hypothesis of the Ebro border formulated about ten years earlier. These last Neanderthals apparently left their traces in the neighbouring Gorham cave. We decide to go with Clive to this other cave. Gorham is only a few dozen metres from the immense cave where we meet. Clive had been researching Vanguard Cave for about thirty years because of its proximity to Gorham Cave, which had been known since the beginning of the twentieth century and had been explored archaeologically since 1948. Clive himself continued his research there from 1999 to 2005. He was convinced that this cave, excavated too early, would not be able to provide an understanding of the precise structure of the Neanderthal occupations. So he shifted his attention to a few dozen metres away, to the other natural cathedral, its twin sister, Vanguard Cave. His hope was that Vanguard, which no one had thought of exploring at the time, could reveal the same Neanderthal settlements as its prestigious neighbour Gorham Cave. He would not be disappointed. From the vast stone mouth of Vanguard there spews a slope of marine sand. The cave is filled with eighteen metres of sand plastered to the bottom of the cave. Eighteen metres recording, almost continuously, Neanderthal settlements expressed, as in the Mandrin cave, over eighty millennia. From the last temperate episode of some one hundred and twenty millennia ago until the Neanderthal extinction, not that of the forty-second millennium but ... twenty-eight thousand years ago? Nobody can at present resolve this thorny question. And, like all thorny questions, the positions of a large part of the scientific community are becoming clear-cut. Sharp. These carbon-14 dates are indeed not easy to manipulate. Gibraltar lies in the most temperate climatic context of the European continent. Here, there's no baroque concerto of *The Four Seasons:* Vivaldi would have had to adapt his work into two seasons, a long hot and dry period followed by a rainy season. Twenty to twenty-five thousand years ago, while Europe was enduring one of the most significant drops in temperature in its climatic history and sea levels were dropping to one hundred and twenty-five metres below their current level, the caves of Gibraltar still recorded the presence of temperate species. During the coldest moments of the last glaciation, when the entire northern hemisphere was largely covered by ice, the olive tree continued to grow on the paths of the Rock. The most marked climatic phases affected this tip of Europe only very secondarily, probably accentuating an aridification of the environments but without the development of the polar environments that impacted on Eurasia and the Americas. These environments, probably favourable to human populations, were not so for the preservation of archaeological bones. And it is the decay of the bones that now prohibits their dating by classic carbon-14 methods. The only way for carbon 14 to overcome these constraints is to establish these dates no longer on bones, but on charcoal. And small charcoal pieces tend to wander, to rise, to descend through archaeological soils. This methodological constraint very directly affects our understanding of the real ages of these archaeological occupations. But the question of a final Neanderthal survival at the edge of the Eurasian world has now been raised and should not be treated too lightly by its detractors. Concrete data, on Gibraltar and across Spain and Portugal, raise the possibility that this immense peninsula hosted the last Neanderthal populations. But the discussion around these last Neanderthals is bitterly controversial. And is it really so important in understanding this historical process that these populations died out all at once, as if they were a single individual, in a deadly wave that suddenly swept across the planet? They could well have died out in a slow agony of humanity. In the end, they died. All of them. And it is these processes of extinction that we have not yet been able to question in their historical reality. As if we could not look death, concrete death and not simply theoretical death, kept at arm's length, straight in the eye. We probably fail to understand the greatest extinction of humanity ever recorded because we can't put one of the deepest taboos of our societies into words. These few taboos, these unconscious prohibitions of thought, systematically group together around what reminds our humanity that it is an animal creature and not a divine creation. Death isn't the only Western taboo. We recognize, alongside death, nudity, defecation and sexuality. These spaces represent our deepest prohibitions. If these prohibitions seem to be rooted in realities very distinct from each other, they nevertheless have one thing in common. These prohibitions, these moments of concealment, deeply and obviously connect our humanity to the natural sphere – to the animal sphere. The same deaths, the same arousals, the same structures of our bodies, are all realities, all self-evident facts, that we cover with a modest veil.

Through these taboos we project humanity outside the animal kingdom.

If the rejection of the ages of the Neanderthal settlements in the Gorham Cave comes with the objective possibility of contesting, discussing and rejecting the proposition of Neanderthal persistence, it also involves a closing of the mind in other ways, ways that reveal other reflexes specific to our societies and perhaps, more deeply, to the mental structures of our *Sapiens* populations. These populations suffer from moving away from common thoughts, dominant thoughts, which represent a perverse unconscious process that prevents us from directly confronting the realities of our universe. We are here faced with a self-blocking process that prevents us from looking at any reality if it's too divergent from a smoothed-out way of thinking. It's annoying, all the same. Avi Loeb, the director of the astrophysics laboratory at Harvard University, clearly lays bare – in his book *Extraterrestrial*[5] – the intellectual process that prevents scientific communities from thinking outside the box. It means that we all think alike – that we cannot think freely. So we go round in circles. All together. These processes of uniformity of thought and terror in the face of the unknown seem quite natural but represent one of the greatest obstacles not only to scientific research but to all forms of thought. And, even among our freest thinkers – and this is perhaps also our shame and our sin – the acceptance of divergent scientific data has never been achieved except in pain, by caesarean section. The reception of divergent thoughts and data then risks tipping over into a question of faith, of belief, rather than into a balanced evaluation of the realities present. The 'why not' does not structure scientific thought on any deep level. In science, we often insist on the fact that an exceptional discovery requires exceptional evidence. We instinctively understand the meaning of this proposition. A discovery that shifts a dominant position towards an uncertain 'elsewhere' can be welcomed only if it is remarkably well supported. This is the least we can expect from the scientific approach. But the construction of our hypotheses, our questions, cannot be limited by the sole use of remarkable data and the rejection of any divergent data, whatever their nature. For an exceptional discovery we require exceptional proof. Avi Loeb noted that, after all, proof is proof. Even if it is divergent. It opens doors, spaces to explore. Divergent thoughts invite us to engage our scientific explorations on side roads. If only in its principle, doubt represents a necessity here. An absolute necessity. Doubt, positive doubt, doubt in the face of the most dominant thought will always represent a door to an elsewhere. An exploratory space that we need to move into fast. We fundamentally need these acceptances of doubt and uncertainty in order to be able to move forward in all disciplines of thought. But *Pourquoi pas?* (*Why not?*) was also the name of Commander Charcot's ship, which sank with all hands – a crew of thirty-nine members – on 16 September 1936 off the coast of Iceland. And this parable reminds us that uncertainty always represents a danger, and that it is better to stay in line so as not to stray too far into uncertain lands abandoned by the entire group. More than being a trait of our Western societies, might not this way of thinking about our universe, all together, sometimes against the reality of the world, be a direct echo of the ethological nature of *Sapiens*? The standardization of all ancient crafts in *Sapiens*, that standardization of our only humanities, which I crudely highlighted in *The Naked Neanderthal*, is not at all tied to the crafts and techniques of our direct ancestors. They are most likely just an echo of something much deeper that structures our human nature. Our way of being in the world. All together. Wouldn't we have, here, the strength and, perhaps, the Achilles heel of our own humanity?

But we're going too fast, too far.

We sit down with Clive in the heart of the cave and begin to discuss the very singular nature of the Neanderthal occupations of Gibraltar. The waves crash at the foot of the cave and the Mediterranean stretches out in front of us, from the cave mouth to the horizon. To the south we can make out the coast of Africa clearly outlined above the sea. Between the two, a dozen enormous merchant ships and oil tankers are slowly moving towards the nearby refineries. The horizon is orange.

'Is the colourful sky caused by pollution?'

'The air's perfectly pure here. It's the sands of the Sahara that the winds bring to the Spanish coast. You see Ludovic, the cave is completely filled with sand deposited by the winds. And we find these orange, oxidized deposits there, which have come from the Sahara. But we don't just find them here, in Vanguard Cave, or in Gorham Cave right next door. The entire eastern side of the cliffs of Gibraltar is filled with these sands. It's the east wind that blows most often and deposits them along the long bar of cliffs of the Rock. The city of Gibraltar is positioned on the other side, on the west, protected from the prevailing sea winds.'

The sandy slope on which we were sitting was only a tiny sample of a much larger reality. Here, on the eastern flank of the Rock of Gibraltar, the winds regularly deposited their sand, filling not only the mouths of these caves facing the Mediterranean but also covering the whole of the eastern cliffs of this immense rock with sea sand. Here there are eighteen metres of sand but along the cliffs there are three hundred metres of sediments flattened by the winds. Palaeolithic sands that cover an unknown number of caves, shelters, cavities. Gorham and Vanguard are only the cavities revealed by the sea that comes to gnaw, whip, absorb, reopen, these immense caves once filled by sand. I begin to understand. Gorham and Vanguard are not two distinct sites located a few dozen metres from each other. These two mouths licked by the sea represent a single site some distance from each other. No, it's the entire cliff of Gibraltar, over its six kilometres of length, that has fossilized the Neanderthal camps, passages, settlements. Gibraltar is an immense time capsule, a machine naturally fossilizing the slightest Neanderthal camp on its paths for eighty millennia. During the great cold of the last Ice

Age, the sea was between eighty and one hundred and twenty-five metres lower than it is today, pushing the Mediterranean shore far out, to five or six kilometres from the caves. And the rise of the sea during the last big thaw, *our* big thaw, eroded the most advanced, easternmost parts of these sandy deposits. If Vanguard was filled to the ceiling some forty millennia ago, Gorham, right next door, continued to record much later human occupations, Phoenician, Iberian, Greek and Roman. In their details, the records differ, but a single process explains these Neanderthal fossilisations as well as their revelation by the rising of the waves. It was the sea that carried away the layers of sand, finally reopening the cavities that sand had closed. It is the opposing battle of the winds and the sea. Winds that deposit and fill, and the sea that gnaws and reopens.

The oases of the sea

It was the rise in sea levels that would allow the reopening of the cave and give access to its archaeological levels. How many of these caves are still buried, invisible, buried by the sands deposited by the winds, from here to the northern tip of the Rock where the skull of Forbes' Quarry was revealed in 1848, not by maritime erosion but by the quarrymen intent on fortifying the British citadels?

Clive's work demonstrates that this group of caves forms a single system. A system of great complexity but which can only be understood in its entirety. A total archaeology covering the entire eastern flank of the Rock of Gibraltar would probably reveal archaeological records that are among the most important fossilized Neanderthal settlements across Eurasia. The whole forms, in its potentialities, an incredible complex that goes far beyond the imagination of a simplified archaeology built on the limits of small trenches skilfully positioned in each of these cavities. Such a system of archaeological records is dizzying but not, strictly speaking, unique. The real potential of a certain number of archaeological sites is commonly underestimated, particularly in contexts linked to the actions of the wind. The wind probably constitutes one of the common points in the fossilization of such mega-groups of Neanderthal sites. Winds can cover, fill and fossilize immense areas. But, faced with such archaeo-logical immensities, an archaeologist's life is not enough and the tools allowing one to confront such areas probably remain to be invented.

This may be surprising, but the enemy of the archaeologist is not rarity, but the management of profusion and the ability to understand human occupations not as sites isolated from each other but as systems. Systems whose hypercomplexity demands to be explored. All well and good – but it's easier to say, to theorize, to put it into words, than to carry this out in the real world. On some of my field trips I had already encountered comparable realities in which one is tempted to track Neanderthals no longer between the two walls of their habitat, in the cave, but over vast expanses, from valley to valley, like a trapper. But the very notion of a habitat is meaningless here. These nomadic populations were fully part of their space and could not have conceived of a notion of habitat that would fit in with our imaginations, our experiences of what habitat is, in our daily lives. The Neanderthal habitat is expressed neither between the walls of a cave nor in the plains, between the stretched skins of their shelters. Those ancient populations, in this case Neanderthals as well as *Sapiens*, in their nomadism, physically belonged to much larger spaces not limited to the perimeter of their homes nor to the spaces they occupied for a few nights. Habitats and territories are intertwined here. Their combination was of structural importance for those populations. But for us archaeologists, these notions are distinct. The different notions of habitat – the nomad's and the archaeologist's – may not describe the same realities, the same concepts, or the same feelings, and the original reality of these notions may lie in intangible spheres that communicate only imperfectly with each other. The habitat of the archaeologist is a construction, an analysis tool. The conception of habitat among those nomadic populations was rooted in the immense territorial expanses in which they experienced their daily lives. And, as I sit in this cave facing these expanses, the opening of thought onto these distant societies as a system becomes obvious. It represents the hope of an understanding no longer based on the mere juxtaposition of archaeological sites. But a slightly crazy hope. A little hazy, too, as the notion of time escapes us here more than elsewhere. To put things into a network, a system, you have to be able to connect points together. And have a more or less precise idea of the time separating or connecting the points together. In theory.

I join Géraldine and Clive's little team that has taken over the archaeological operations. They are slowly digging in the sands, pure sands in which, less than a metre thick, dozens of dark lines streak the sediments. Dark, almost horizontal lines highlight and punctuate the yellow sands. These lines are the traces of Neanderthal fires. Successions of fires. The diggers have just cut out one of these hearths. We can make out a beautiful flint point placed on the edge of the hearth. They are kind enough to extract it, and place it in my hand. They await the verdict of my gaze. This expertise is not interchangeable. The gaze of the one who has seen is awaited – the gaze of the one who has directly seen millions of these crafts throughout the ancient world. Knowing that I will share a few moments with them, the team was also kind enough to bring back to the cave a collection of flints from these archaeological levels. Remains abandoned some seventy millennia ago. These are superb objects. They are disturbingly fresh, as if they had just been carved and abandoned in the sands. The sight of these very clear alternations of black lines and yellow lines punctuating the ground informed me immediately that everything here had been frozen in place. After the departure of the Neanderthals the remains had obviously been quickly covered by the sands. Their only movement, their only mobilization for seventy thousand years, depended on the hands of archaeologists waking up this sleeping beauty. The collection is magnificent. But, better still, it's surprisingly familiar to me. It's not just that these technologies alone remind me of technologies seen elsewhere. It's deeper. It's more subtle. It's the style. The manner. Something that goes far beyond technologies and craft knowledge.

'Are these objects common in your archaeological records?'
'The selection is representative of what we find here, Ludovic.'
'Excellent. I'm really happy to be able to hold such objects. But it's a lengthy sequence. Your 18 metres of archaeological records cover 80 millennia. From 120,000 years to 40,000 years ago. This is precisely the range over which we are investigating at the Mandrin cave. In our country, these 80 millennia are only recorded in three metres of sand. But in these three metres, I recognize remarkably diverse craft traditions. I think that we can distinguish in the Rhône Valley something like eight major cultural phases, which would indicate the existence of profound cultural upheavals every ten millennia on average. Finally, in detailed terms we must have moments of stasis, of population stability, particularly in the ancient phases, then in the last

twelve millennia everything accelerates. As if everything was now changing at a frantic pace. Anyway, a frantic pace, I hear myself, there are five major recognizable cultural phases, very pronounced, very marked, over a dozen millennia. It's a frantic pace over periods of 2,000 to 3,000 years …'

Bursts of laughter. We're really starting to swap ideas.

'We don't see anything like that here, Ludovic. Our flints from the 120th millennium seem perfectly identical to those from the 40th millennium.'

Mutual silence.

'Ah … that's fundamental. That's really interesting. This doesn't match anything I see in the Rhône Valley. And it opens up some great horizons. Well done to the whole team. Great job.'

I hand the flints back to the teams and go back to sit next to Clive. This homogeneity, this form of stability in technological traditions, seems to structure all the Neanderthal settlements recognized in Gibraltar. A very long durability in knowledge and ways of being. It's here the notion of time that we can question. I noted that the notion of time in Gibraltar escapes us here more than elsewhere. By that I mean that it is difficult to certify that such and such an archaeological unit comes precisely from such and such a millennium. Here the fragility of carbon 14 does not help to simplify such an equation. How then, in this hypercomplex system of the different Neanderthal marks of the Rock of Gibraltar, can we network the information from these different cavities? How here, in this environment so rich in Neanderthal occupations, can we join up the points? But if societies show real static periods over eighty millennia they also free themselves a little from the notion of time. Not that this notion isn't essential to us as archaeologists but the overall understanding of these Neanderthal organizations then becomes possible as a system. If these societies are understood as a system, their organization doesn't necessarily obey temporal constraints as they appear to us, and are essential to us a posteriori. These temporal permanences, like the notions of habitat and territory that we have touched on, find no echo in what we ourselves experience in our understanding of the world. The instability of our sedentary societies seems punctuated by time. Centuries and millennia inflict their modifications. Time can then only be perceived as a fundamental factor in our historical mutations. The Neanderthal permanences that we encounter here at the tip of Europe do not, however, represent an anomaly in these archaeological records. It has often been said of these permanences that they constitute a fundamental face of the organization of these populations. Should we consider, faced with such realities, the existence of achronic human populations? Ahistorical populations, eternally the same? If this has often been suggested concerning Neanderthals, this observation was for me unexpected and particularly exotic, corresponding to nothing that I knew through my own research. In the Rhône Valley, societies mutate, change, replace themselves again and again. Nothing that could resemble a quasi-eternal stability in the organization of these astonishing creatures. For me, Neanderthal societies evolved on temporal steps quite comparable to those that we can diagnose elsewhere within the *Sapiens* populations of the Palaeolithic. But this view was the Rhône view. A singular space that perhaps truncated my own vision of the possibility of other realities, in these societies, and in other places.

If such ahistorical societies could have existed, such realities directly question the way in which we could question them. Temporality would then mainly represent, in these immutable contexts, a tool, an essential tool for organizing our own analyses, our own correlations but it would become a much more subsidiary tool in the reality experienced by those populations in their own time. Here the societies of the one hundred and twentieth millennium deeply resembled, in their knowledge, in their traditions and in their organizations, the very last Neanderthal societies, whether the extinction took place twenty-eight or forty-two thousand years ago. While this leap of fourteen millennia affects a certain historical reality of these populations in relation to the external world surrounding them, it perhaps had no impact on the structure of their internal organizations. At least in what we are able to perceive of it. But archaeological records are often very complete. We see a lot. The permanence of these human societies is probably not a simple optical illusion in the face of populations of whose social and cultural realities we can perceive only a tiny part.

I sit down again next to Clive, who has expressed his desire to share with me his view of the organization of the Neanderthals of Gibraltar. We silently look towards the sea. I had not imagined such Mediterranean proximity. There's a whole herd of dolphins swimming a few dozen metres away. The television crews continue to film the archaeological research and we take advantage of it with Clive to exchange our views. We are in his place, in the spaces that he has been exploring for three decades and I let him express his knowledge, positioning myself to listen in the remarkable setting of this immense cave.

'Ludovic, our Neanderthal levels give us the whole procession of the resources of land and sea. We have remains of deer and horses, but also dolphins, walruses, penguins, fish, molluscs. But the sea at that time did not come to expire as it does now at the edge of the cave. One hundred and twenty thousand years ago the waves entered deeper into the cave. But during the last Ice Age, and for tens of millennia until the Neanderthal extinction, the coast was positioned about five kilometres from there, approximately where those large cargo ships are. On the plateau that's now underwater, which spread out in front of the cave when the Neanderthals occupied it, we have identified, using underwater explorations, areas in which bubbles rise from the depths. These are resurgences. Sources. In the small space of this five-kilometre plateau that developed between the cave and the sea shore, there were rivers, but also depressions that hosted lakes. These populations benefited from an amazingly diverse biotope. And the great ice ages did not have a direct impact on this region, which was able to function as a refuge for all the animal species that were adapted to temperate environments and could no longer occupy the highest latitudes of a Europe under ice. But you'll have seen that in the studies we've published.'

The climatic, environmental, biological particularities expressed by Clive framed his vision of a late persistence of Neanderthal populations on the tip of Europe. It was not only the dating of Gorham's cave towards the twenty-eighth millennium that structured Clive's thinking but a much richer vision that repositioned the roots of these Neanderthal populations in unique ecosystems. Oases of the sea.

Where three paths of the Mediterranean meet.

'Ludovic, I wanted above all to give you two important pieces of information that I wouldn't want to reveal to you directly in front of the cameras. We haven't published it yet, but we've resumed field studies in recent years on the Forbes' Quarry cave in the hope of finding the level from which this Neanderthal skull comes. We think we've located it and we've set up luminescence measurements in the sediments in order to determine the age of this archaeological level. We've obtained some new results that were quite annoying for us because they're not compatible with the results of palaeo-genetics which indicate an ancient age between 70,000 and 120,000 years. Our measurements constrain the age of this skull to a timeline between the 60th and 40th millennium. And most likely between 40,000 and 50,000 years. This corresponds precisely to the details of the studies you sent me on Thorin which imply a profound reconsideration of the age of Forbes' Quarry. Your team's studies from the Rhône Valley and ours on Gibraltar allow us to show that Thorin and Forbes lived at precisely the same time. The biological connection that you demonstrate in your studies, Ludovic, has also deeply affected me. You know that we modelled the landscape as it was 40 to 50 millennia ago. When we lower the sea levels by 80 metres we see this plain in front of the cave. This plain is in reality a vast continuous corridor which directly connects Gibraltar to the Rhône Valley.'

I remain silent. On the one hand, I'm not sure that I've understood the ages that he has just expressed to me, and above all I don't want to interrupt the flow of this speech. I know that this is one of those moments that ethnologists and sociologists have widely described. Moments when speech is freed and when information, sometimes long held back, pours out naturally. We must learn to recognize these moments, these parentheses where the exchange of ideas must above all consist in listening, so as not to break the continuity. They're always like an actor's asides. My doubt about the understanding of these ages will be resolved a little later, when Clive expresses in front of the cameras what he had just expressed to me. This will ensure that my surprise in front of the cameras is not feigned. And I will have him repeat it to make sure that I've understood what he is transmitting to us in real time.

The man from Mandrin and the woman from Gibraltar did indeed live at the same time. These data definitively and profoundly impact on our understanding of the last Neanderthal societies. A population structure therefore exists, indisputably, within the last Neanderthal populations and in the very heart of Europe. The Thorin line, separated from the classic Neanderthal line, persists in neighbouring, connected territories for an almost unimaginable temporal thickness. Sixty millennia. Sixty millennia during which the two Neanderthal populations will diverge profoundly. Sixty millennia without any exchange of genes. A flick. Just think, this temporal power is more important than that which allowed the wolf and the poodle to diverge. All is well. A major element of the equation describing the Neanderthal extinction has just been revealed. It is no longer just a suggestion: it's now been powerfully demonstrated. What are the implications?

The classic Neanderthal lineage is recognized throughout Western Europe and can be traced back nearly eighty millennia, hence the suggestion of a great biological homogeneity of Neanderthal populations until their extinction. This lineage is not limited to Spain, Belgium or Germany. It is attested in regions neighbouring the Mandrin cave, in Charente for example and in very close temporalities. But the classic lineage and the Thorin lineage diverged and did not exchange after the hundred and fifth millennium – after that millennium, in a mild climate, to the Neanderthal extinction during the last Ice Age. The populations diverge. The divergences revealed by genetics fit into a landscape of knowledge about the organization of these populations. The organization of Neanderthal societies is well known in the classic regions of prehistory, in Dordogne, Charente, Burgundy. The understanding of these Neanderthal societies was extended greatly in the twentieth century in research by the teams of François Bordes and the Bordeaux school; concerning the final Neanderthal phases and the question of the Châtelperronian, it was the teams of André Leroi-Gourhan in northern Burgundy who defined the intellectual frameworks that still profoundly affect our understanding of these societies. Following them, it will soon be three generations of researchers that have been exploring and defining the technological, cultural and social structures of these populations since the 1980s. My own research on Thorin developed more than ten years after the establishment of the cultural organizations of this Neanderthal society. While Thorin was discovered in 2015, the cultural structures of these Neanderthal populations were described much earlier, in 2004, in my doctoral thesis. This work highlighted the profound originality of the technological traditions of these populations and proposed the recognition of original cultural frameworks that absolutely did not coincide with those of the classical regions of the rest of France. I thus no longer defined divergent technological traditions but proposed the existence of deeply differentiated cultures – cultural provinces expressed over a very long period between the Rhône Valley and the classical regions of prehistory, Dordogne, Charente and Burgundy, leaving many of my colleagues uncertain as to the meaning of my hypotheses. This is how the Neronian was born, which we now know corresponds to the oldest *Sapiens* incursion into the Neanderthal territories of the European continent. The Neronian was then replaced by more classical Neanderthal traditions, but these still did not fit with what was known in the neighbouring regions. The post-Neronian I and the post-Neronian II were born. If their name simply referred to their position in time, after the Neronian the need to create specific names for these archaeological groups arose from the impossibility of finding obvious connections with the Neanderthal societies classically recognized in the Atlantic area and as far as Burgundy. To put it one way, for me there was then a classical Neanderthal region, a Neanderthal biface region and a Rhône area, where the biface disappeared in very ancient times but where we find technologically remarkable flint points. And in my eyes the Neanderthal craft traditions of the classical region were particularly exotic. They did not fit with anything I knew. While the quality of the 864 pages of my study was unanimously praised, its conclusions highlighting deeply differentiated cultural provinces were obviously more difficult to accept as these conclusions went against the grain of a rather rigid set of ideas regarding the cultural structure of Neanderthal populations.

And Thorin is now part of this history of research. He does not just unexpectedly reveal the existence of profound biological divergences within Neanderthal populations. He is part of a history of research that had long theorized the possibility of such a divergence without us being able to understand the profound meaning of these technological and cultural divergences.

The matter is settled: it's still impossible to make the interpretative leap between biological populations and cultural populations. Where are the Neanderthals? Where are the *Sapiens*? Who were the authors of the Châtelperronian or of these crafts that we perceive as Neanderthal in Gibraltar or on the polar circle? So many unknowns in the equation, blurring our understanding of this process of extinction of humanity. How do humans die? And how are we to conceive the processes leading ultimately to the greatest extinction of humanity ever recorded?

But the astonishing trans-Mediterranean connection linking Gibraltar to Mandrin perhaps spoke to us of something other than simple genetic relationships, distant even though they crossed space and time. In Gibraltar, the teams noted continuities of craft traditions over more than eighty millennia. Continuities that are recorded as eternities. It is very exotic compared to what I see in the Rhône Valley, where traditions are overturned radically and always abruptly, without slow modifications, without slow evolutions like those I encountered in Ksar Akil on the other side of the Mediterranean. On these three Mediterranean shores I could see clearly differentiated processes.

On the easternmost shores of the Mediterranean in the old strata of *Sapiens* from the fiftieth to the fortieth millennium I held in my hands objects that spoke to me of very gradual mutations. As if the techniques were biological entities – living entities evolving slowly, very gradually through the millennia, to mutate into something else.

In the far western Mediterranean I noted perpetuities, reproductions through the millennia. Without change. Without evolution. A total stability that would never be undermined until the biological extinction of these humanities. No slow evolutions. No mutations towards something else, either in ancient times or in the final moments of Neanderthal societies.

Between the two, in the immense Rhône valley I had a third pattern totally independent of the historical structures discernible at both ends of the Mediterranean. The Rhône valley recorded abrupt and profound cultural changes. The Neronian *Sapiens*, the post-Neronian I and the post-Neronian II carried out by Thorin's people. Clear-cut technological and cultural realities. This was neither the slow eastern mutations nor the incredible stasis of the West.

These processes express very distinct historical realities that can be tracked in three disjointed Mediterranean spaces but whose stories end up, in a totally unexpected way, converging at the very heart of the Rhône valley. Mandrin is located 3,000 kilometres from the flanks of Mount Lebanon and nearly one thousand three hundred kilometres from

Gibraltar. The Rhône Valley represents a geographical in-between and a third way in the evolution of human societies. An original path but one that could well bear witness to the meeting of the two historical processes that can be discerned in the east and the west of the Mediterranean.

In Gibraltar, the ecosystems were never impacted by climate change or by the possibility of any food shortage. Mussels and oysters, fish and mammals were all consumed, and constituted a general framework in which these populations had never been able to experience the slightest serious environmental pressure. No shortage had ever occurred here. And yet, at the end of the race, even in these exuberant oases of the sea the Neanderthals died. Whether the extinction took place during the forty-second millennium or in the twenty-eighth, the Neanderthals died out. Here later than elsewhere, as the model of the Ebro border would have it? Radiometric measurements and dating do not allow us to decide too abruptly without taking a largely subjective position. But we could well have the last Neanderthals here.

The structure of such a model does not rest as much on the richness of the biotope and the mildness of the climate as on the exclusion from these areas of the *Sapiens* populations that seem to be attested here only in much later phases of the Palaeolithic. Here, as elsewhere, *Sapiens* remains at the heart of the equation of the extinction of a certain humanity. And it was neither the favourable climate nor the richness of the biotope that would be at stake, since these two conditions are not specific to Gibraltar but remain widely shared throughout the Mediterranean rim, without these conditions altering the finality of this extinction of humanity.

In the end, the Neanderthals die.

By broadening our sphere of analysis, we note that in the south of the Italian peninsula, at the southern tip of the Italian boot, in very comparable Mediterranean climates and faced with biotopes presenting the same balance as in Gibraltar, the Neanderthal extinction could indeed have been expressed in quite ancient times. These extinctions could well have taken place here between the forty-fifth and forty-second millennium. And in Italy, on the other peninsula of the western Mediterranean, several studies now demonstrate that *Sapiens* arrived very early, probably as early as the forty-fifth millennium.

While the climate and the richness of the biotopes seemed to be essential reasons for the Neanderthals not to die out on Gibraltar, the combination of the elements of the equation allows, here as elsewhere, the reasons for the greatest extinction of humanity to coincide with the phases of expansion of modern populations across the continent. These facts then suggest that, if there is persistence on the westernmost peninsula of the immensity of Eurasia, the process was probably not correlated with favourable biological or climatic realities. Such a temporal transgression may well not be due to the mildness of the biotope, since this remained constant even during the greatest cold of the Ice Age maxima. The main factor here was the geographical distance of this Mediterranean area from the rest of the old world.

And the Rhône Valley seems to have borrowed from these two distant cultural universes of the East and the West. Mandrin would record both the first *Sapiens* migration to the west of the world, connecting very early with the Mediterranean East and, subsequently, the migration of new Neanderthal populations from the Mediterranean west. In both cases, coastal routes seem to have been at play in these connections, whether these connections resulted from purely maritime movements or whether the populations took the long coastal corridor that sea level models have highlighted. The existence of such a geographical corridor is not insignificant, since the eastern flanks of the Iberian Peninsula today appear as a succession of mountain ranges that can exceed three thousand metres in altitude. But these routes, perhaps submerged by the waves, remain unknown to us. Whatever paths these humanities took, let us keep in mind that this observation is now certain. The last Neanderthal populations of the Mandrin cave were, biologically, those of the southern tip of Europe. And these Neanderthals separated out from all the other populations of Western Europe a very long time ago: genetics reveals the existence of unexpected human and biological boundaries. These were total separations, distinguishing Neanderthal populations for tens of millennia. Separations that the analysis of crafts had already noted but whose texture we could not define. Mere cultural divergences? No. Separations of humanities, divergences and the abandonment of all exchanges.

What an amazing story. And do we not have in this unexpected discovery certain elements fundamental to our understanding of the greatest extinction of humanity? We can see distinct historical processes emerging between each point of the Mediterranean. These profound millennial divergences finally come together in the Mediterranean artery of the Rhône. There's the first *Sapiens* migration from the Mediterranean Levant in the fifty-fourth millennium. Then the return of Mediterranean Neanderthal populations along the great river. The return of the peoples of Gibraltar northward from the western Mediterranean. A return from the oases of the sea. From those Mediterranean slopes whose richness and mildness of biotopes have no equivalent in continental Europe. I briefly mentioned the flint crafts of Vanguard and Gorham, because here a lot of work still needs to be done and there are decades of research awaiting shared perspectives. But these Neanderthal tools, in my hand, expressed a deep familiarity, in which genetics told me that I was confronted with precisely the same humanity. It is this familiarity that I have never been able to find in the classic Neanderthal crafts of the Atlantic regions. I have long been expressing this divergence, since well before the new prospects offered by genetics, to my colleagues from Dordogne, Burgundy and Charentes, whose Neanderthal tools seem so exotic to me. There's exoticism on the one side and familiarity on the other. Human barriers. Genetic barriers. And those invisible borders, those borders of humanity of which we knew nothing.

In the Rhône Valley, the pulse of human migration ran aground for a time. Coming up through the Mediterranean from its east and then coming up again, but this time from its west. And from the north, through the Alps. These pulses are the cultural divergences that I have been documenting for thirty years. Well, ultimately I was wrong. The Mandrin cave may not express the rapid evolution of human traditions in a geographical space but gathered, not the influences, but the different incursions of populations that were rooted from all eternity in other geographical spaces. What we see in Mandrin are the tails of comets. The passages, for a time, of societies that had structured themselves elsewhere.

I here have a diagram that explains the three historical Mediterranean realities. The slow eastern mutations and their incursions towards the Mediterranean west. The perennial features of Gibraltar, a true reservoir of populations in their seaside oases and the recolonization of the intermediate spaces in the centre of the Mediterranean after the departure of the first *Sapiens*.

And a final bouquet, a full stop adorning the millennia of this maritime waltz: extinction.

The end of the end. The Punta de Europa as the last stopover of the *Sapiens* populations. Once the Strait of Gibraltar is reached by *Sapiens*, the loop is finally closed. That's it.

From the two shores of the old world that contemplate each other here from Eurasia to Africa, it is now the gaze of *Sapiens* alone that projects itself over all the continental expanses.

A simplified world

And *Sapiens* lies at the centre of the equation. Again. There was no huge volcanic explosion here. No decline of populations faced with the collapse of their biotopes. No stellar radiation eradicating human populations. No deadly epidemics. No, nothing at all that can construct a discourse on material, biological, climatic contingencies that are, above all, disembodied – as we cannot be responsible for the greatest extinction of humanity.

But there you have it: putting the advance of *Sapiens* at the centre of the equation does nothing to resolve the enigma of the extinction of these populations. How did the last Neanderthal societies die out? And within these populations, how did the last Neanderthal die? In a slow agony, until the death of the last of the living, all alone? Brutal extinctions, en masse? No one knows how humans die. Nor how our humanity will end, now it's united in a simplified world in which ancient humanities have disappeared.

A humanity now united?

Yet we perceive this humanity, *our* humanity, in the present moment, in a wealth of cultures, traditions, languages, ways of being in the world. These perceptions, these constructions, feed into all tensions, all wars, all annihilations. And, while in our daily lives the great story seems to invite itself once again to the great meal of the dead, we ultimately differentiate ourselves when it comes to the political necessities of dividing our horizons. This humanity, ours, nevertheless simply represents different faces of a single, unique way of being in the world, united not only in its biology but more deeply, under the veil of our constructions in the entirety of its deepest values. In 1985, Sting spoke to us about the Russians, the hysteria of our disunited societies, a hand, a mind, a hope stretched towards the East announcing the glimmers of the Soviet collapse a few years later. Has the post-Soviet parenthesis closed today? Could our own history of East and West, of exchanges and closures, be a distant reflection of those old stories of the fortieth millennium and beyond? Can East and West be defined in short pulsations of exchanges, interspersed with long static periods when populations are separate? While millennia pass in hundreds of moments of uncertainty? Everything here needs to be explored.

But make no mistake. Here and now everything is simple. We are in a simplified world compared to what it was in its history, when biologically differentiated human forms crossed paths or ignored each other across the millennia. When humans still possessed a diversity that was not basically cultural but purely biological. When humans were also part of the same diversities that we recognize in the animal kingdom. There was no simplification of populations outside of humanity. Dozens of forms of canids, dozens of forms of swallows. One humanity. So the world of humans is now a simplified universe. Simplified by its past extinctions.

Is this simplification, a selective anomaly that rather crudely distinguishes our humanity among living creatures, natural? Or could it be artificial?

Today I wonder how humans die?

3

How do humans die?

A grammar between two worlds

What does this post-Neanderthal world look like, which seems to be such a focus of attention?

Among the old Neanderthal societies tradition is serious business. Imagine that the broad frameworks of their crafts, still widespread across Eurasia in the forty-second millennium, are inscribed in the same techno-logical structures as those that could be recognized in their ancestors two hundred and fifty millennia ago. Not that these populations have no tradition, or are frozen as if in evolutionary stasis. No, no, not at all – but the fundamental structures of their organizations don't seem to have undergone any major upheavals during this immense temporal thickness. There's a well-established atmosphere that reproduces itself through the ages without breaking their ancestral balances. We can perfectly well recognize traditions, ways of doing things, styles, which last for a time, then are replaced by other styles, other ways, but which are always part of a certain way of being in the world. With Neanderthals, let's not wipe the slate clean and erase the past. But how can we define these amazing properties? In reality, we're not simply talking here about technological systems, about this or that way of knapping flint, of making tools, of making weapons. It's probably something deeper that can probably be located in a layer underlying the very structure of these technological systems. A more subtle layer, anchored in a certain background. It's not a question of flint points or flakes or blades or this or that type of object or the presence of this or that technology. It is, rather, ways of being that are reproduced over hundreds of millennia. But it's still difficult to describe these ways of being in plain and simple terms. Too many constraints come into play. I use the phrase 'ways of being' as I might talk of style or phraseology or grammar. Yes, that's more like it. The set of rules that enable the whole to assume a certain meaning. A certain semantic orientation. There's no point here in overwhelming you with technical terms because, fundamentally, in what I'm trying to express, there's almost no advantage in knowing whether the way of making these stone tools is part of the discoid or Levallois kind of carving or something else. It's not a question of technology. They could have made seaplanes and that would not fundamentally change anything in what I'm trying to determine here. The important thing isn't simply to note that these societies endlessly reproduce ancestral knowledge whose origin is immemorial but to note that this borrowing from ancestors establishes balances throughout the ages that vary but always obey the same grammar. Through these tools we can also recognize real traditions. Real cultures. Trace their origins, structures, territories and future. Build up a history of ancient societies, while noting that these various arrangements of knowledge occur endlessly but are always based on certain ways of being. Certain grammars. Not that these Neanderthal populations are incapable of freeing themselves from these grammars but rather that perhaps every human population obeys certain grammars specific to it. And it is perhaps in these ways of being, in these human ethologies, that I already noted in *The Naked Neanderthal,* that something needs to be understood. From this perspective, Neanderthals would not be more fixed or more prisoners of their ethology, of their structural, instinctive, natural, conditioning behaviours, than we are ourselves. But these two grammars are perhaps not the same. When I note a form of singular creativity specific to Neanderthal societies, I also highlight, as if in a reflection, certain singular features of *Sapiens*. A recurrence and redundancy in craft traditions. A phraseology of the repetition of gestures that highlights much deeper structures within our own societies. Within our humanity. Much more deeply buried, too. In a mirror image, in the infinite creativity of Neanderthal artisans creating an edge, shaping a knife or a weapon, a certain relationship with materials is played out. An astonishing dialectic with rocks, textures, colours, shapes. They don't impose on the material. They play, they come to terms with it, and they register their project according to the properties of the natural rock that they transform. Is it simply that Neanderthal artisans adapt to the rocks? Or more profoundly, in this way of not imposing themselves too directly on the material, in this way of playing with what the natural object is, do they not already express, in their gestures, the transformation of their artisanal project? And in this way don't they transform themselves according to the subtle properties in which their project is expressed?

As if the material were not just the passive subject of the artisanal project but the central actor in creation.

As if the material had its reasons, its logics with which one has to come to terms. Rather as if the rock itself were moved by a will of its own. A will that the artisan deciphers, emphasizes, brings out. Perhaps here human hands transform stones and the stones in return shape the humans, in their project. In their will. And this shows the immense creativity of Neanderthal daily life. But it's a technological creativity that blends into the natural environment, into the realities with which the artisan comes to terms.

But on a second reading, doesn't this dialectic speak to us ultimately about ourselves?

In the amazement we may feel when faced with these objects and with this almost infinite creativity that plays with every form, in every way, as if the rock had its own life, its own will. Perhaps my amazement, which I try to share, which I try to verbalize, but which, you see, easily defies words, has a direct connection to what we are? *Sapiens*. To what all our societies are. Like a dividing line. A common point beyond all our cultural differences. In our ways of repeating, always, our artisanal projects, whatever the singular properties of the materials encountered.

And these *Sapiens* crafts, of all ages, of all cultures, of all traditions, evoke other ways of being in the world. Ways that are highlighted precisely because those of the Neanderthals are so different from them, diverge so dramatically, express something else. Crafts obviously don't tell us just about objects or about the artisan's projects. They obviously tell us about the societies that foster such projects. They tell us about the humanities that carry them. About their views. About their grammars. And in these flints from after the fortieth millennium – or well before, if we hang around Africa or the Mediterranean Levant, where *Sapiens* have been wandering for as long as the first European Neanderthals – other balances are evident. And these artisanal redundancies, to varying degrees, these flints that are always the same overall, speak to us about ourselves. About our conceptions. About our rigidities. About our social reproductions. About our ways of being in the world. About our structures. Not as Westerners. But as humans. As the last surviving branch of all other past humanities. And if this proposition, this balancing act, which tells us as much about those creatures as about Us, highlights a grain of truth, there's perhaps a fundamental lesson to be learned from it. Something to reflect on. To express. To decipher. As if the infinite repetition of these *Sapiens* flints suggested the existence of deep structures that speak of our humanity. Of our inextinguishable need to always want to be together, to always want to reproduce the acts, the ways of being of the members of our group.

Apology for the stifled

We harbour this desire to do what the others do. To do the same thing in common. To dress in the same way and in these ways to express our attachment to certain values, our attachment to certain communities, to certain social strata. Lévi-Strauss noted that societies were not distinguished so much by their isolation, their distance in space and time, their singular history, as by their proximity. There's a need to express the fact that we are different from those opposite. It's necessary to emphasize excessively that within our community we make up a society, a unit, and that as such we form a homogeneous mass. A mass distinct from other human masses.

Expressing difference therefore amounts to expressing that we're all identical, obviously. Amongst ourselves. And this need to express our singularity through the expression of unified, codified values, visible at first glance, already formulates this same way of being in the world that we see in the *Sapiens* flints, even going back over several tens of thousands of years. It will be up to us archaeologists, one day, to question its emergence, not in time, but in the very structure of these modes of conception. To seek the sources of these ways, of our ways of being in the world. As for their most basic temporal origins, they appear to be distant, probably well before the hundredth millennium.

And, coming back to ourselves, would it not emerge from these observations, from these propositions, from this need to be together and identical that in *Sapiens* the difference oscillates between something challenging and something quite unacceptable. If the difference is seen, if it is spotted, it is condemned, immediately and profoundly. No one wants to be different. As if a recognized difference induced, like a reflex act, a sentence of rejection. A physical rejection, as in the exclusions explored by Michel Foucault in the asylums and isolation wards of the world. There, the eccentrics, the mad, those unable to integrate, all those who must be distrusted, who must be kept away and reduced to non-existence, are sealed away. They are forced to be silent. Their very existence is denied. Finally, they can't be seen. They can't disturb anyone. Finally, they no longer exist as a daily reality. Gentler, but even more insidious, is moral rejection. In different traditional societies individuals who have transgressed prohibitions are also erased. And if they are not physically excluded and rejected, their disappearance is established by the erasure of their name. It is forever forbidden to utter it. The individual has only an existence of flesh and blood but no longer exists in the words of the group. As Foucault sees it, in all these cases the divergent is ostracized, erased, sullied, denied. The most terrible of sentences. The divergent is extirpated from society, made invisible, put to death physically or morally.

Order must reign, smoothly.

Order reigns in the fear of being perceived in one's difference. The fear of being spotted. The fear that the group will understand that I'm different, since in some way we could all be different. An element of conscious or unconscious fear then nestles in each of us. Like a guardrail. A warning that keeps us in line in spite of ourselves, without the members of the group even having to express the slightest word, the slightest warning. The fear of our difference, of the discovery by the other of our difference, keeps us in line. But the malaise isn't universal: this, our intimate grammar, is the grammar of *Sapiens*.

Boris Cyrulnik also challenges us with regard to this voluntary servitude,[1] the servitude that reminds us that thinking for oneself always leads to isolation. He questions the incomprehensible fact that even in the face of the most abject acts with which the twentieth century so abundantly inundated us the individual preferred to follow the group rather than detach himself or herself from it, rather than express a freedom of conception. How could people choose even when they had freedom of choice to murder young children rather than differentiate themselves from the group? Rather than ostracize themselves? Rather than be divergent? We can murder innocence gratuitously. For nothing. And with complete freedom of our actions. For the sole certainty of being like the others. With the others. Well beyond the nauseating abjection the reality described by Cyrulnik is terrifying: he shows how only a tiny minority always refuses the order that makes no sense. And refuses it not only because the order, the dominant thought, is a vile aberration, but because they cannot follow it simply because it stifles them. Entirely. But the stifled are a very small fraction. A trivial fraction of the social body. To reject the crowd, the mass, a dominant thought is in reality a terrifying act. It lays us bare, forces us to take responsibility for ourselves, as what we are and not as what the group deems acceptable in its ways of being in the world.

Domination takes root here. It takes root in fear, visceral fear, and freezes any idea of freedom. Of any form of freedom. 'Fear is the mind-killer', Frank Herbert reminds us in *Dune*.[2] But beyond fear these overtures point towards Bergson's question about obedience: 'Why did we obey? The question hardly occurred to us. We had formed the habit of deferring to our parents and teachers.'[3] In reality, obedience is unfounded, it is deep, a structural part of our natures, of the social group as a whole that self-controls, self-constrains. 'We did not fully realize this but behind our parents and our teachers we had an inkling of some enormous, or rather some shadowy, thing that exerted pressure on us through them. Later we would say it was society.'[4] And what if the power, the strength of *Sapiens* lay here? What if the incredible efficiency of our societies lay in this way of becoming one body? Of no longer being oneself but a part of the whole? A part of a whole that becomes, through the erasure of the individual, infinitely more powerful?

So would the efficiency of *Sapiens* ultimately be just that? A collective power based on fear? Not a surplus of wisdom but an overflow of the group onto the individual. An erasure of the individual in favour of collective identities.

And we instinctively perceive to what extent a unified social body, excluding in its very nature any divergence, can be profoundly efficient, with a formidable efficiency, total, blind, that always advances as a unified entity. But then, who are the stifled people that Cyrulnik describes? Those who cannot follow, obey in the mass? Are they the first actors of changes in our human societies? Or the first victims of the mass movement, erased and forgotten and who, perhaps one day, will be exemplified when society has changed, at its natural pace – when the new society will need to recognize itself, to justify itself, to legitimize itself in other names, other values, other ways, other models.

And here we are faced with an archaeology of thought in which strata of different humanities are distinguished. No longer by aligning knives against flint blades, not by measuring the number of adornments or the absence of adornments in different people, not by searching for the oldest of the first symbols but by looking beneath the surface in the mental structures of past and present humanities. And if ultimately the picture for *Sapiens* isn't all that positive, it does lay bare the source of the species' remarkable efficiency: it walks along as a unity. Are we still in archaeology or have we already transgressed the boundaries of thought, dragging our gaze towards philosophy? Are we already in the unthinkable? In the inevitable approach of forbidden spaces that are no longer science, real science, the science that counts, that measures, that aligns with an ever more rigid exactitude? We probably have something to learn, to meditate on, under the Neanderthal light, about what we are in our own right.

And this unverbalizable encounter between *Sapiens* and Neanderthals could allow us to better understand our own humanity but it always escapes us. Like the wild men, the yetis and the Barmanous that some still hunt and who become even more intangible when we seek them out. Would the encounter of our humanity with another humanity be linked, in our thoughts, to the same mythological structures? And what promise does this insoluble, unknown encounter offer us?

Perhaps Neanderthal traditions suddenly mutated through the proximity of the *Sapiens* populations settling on their territories, without any precise contact being detectable, since nowhere can the real contact between these two populations be precisely defined, or even simply recorded. We are talking about the acculturation of this other humanity. But an acculturation in the broad sense, a bit as if the simple sight of *Sapiens* objects in the distance was enough to tip all Neanderthal societies into another technological universe. Into some ill-defined 'other thing', albeit one that we conceive of as fully modern since we find in it all the *Sapiens* ways of being after the fall of the Neanderthals. An 'other thing' that would encompass the emergence among Neanderthals of the first manufactured adornments, made of bones, ivory, teeth and all the technologies making use of these hard animal materials. Far beyond that, the emergence of codified communications would be at stake – the surplus of irrational meaning that we put into adornment, clothing, representation, all the frills that have no meaning for those unfamiliar with the codes but whose purpose is to be able to distinguish by appearances and at first glance the status of the members of the community.

You see? No? But look, you *will* understand! No one escapes the sphere.

No one escapes the sphere

The processes of mutation of human societies are still very poorly understood. We don't know how two traditional societies react when they meet. How two visions, two conceptions of the world, two distinct and watertight imaginations react if we place them in contact. Collapse? Creolization? Acculturation?

Can two fundamentally different societies really meet? I mean yes, of course, bodies or eyes can meet, physically. But minds? Traditions? Conceptions? Mythologies? And when I talk about mythologies, I'm not talking about Odin hanging by one foot from Yggdrasil, the world tree. I mean the fact that no society interacts with the objectified reality of the world. I mean that the reality of the world doesn't exist. It's annoying, but that's how it is. Our view, our understanding, our analysis, our reactions, our interactions, our interpretations and our conceptions are only representations. Constructions. Our perceptions of the world around us are totally constructed. They are not realities at all but non-objectified constructions. There is no other reality than that of the social constructions that inhabit us totally. This is what I call, here, our myths. Our mythical constructions that allow us to conceive and interact with what we conceive as external reality, the one that our senses report to us.

Make no mistake, none of this exists as we conceive it outside of our own representations. It's our mental universe or rather our sphere. And through its history over the millennia each society has built its own mental sphere. Each society can be perceived as a sphere. Each society has its own sphere of representations. These representations are not intellectualized; they are not conscious. We absorb them from the time we spend in our mother's womb. And we do not construct them. They construct us. They construct our views and capacities to see and perceive in our sphere and prohibit us from perceiving anything beyond our sphere. They constrain our imaginations about the state of reality and what might exist beyond our own bubble.

You see? No? Okay. We need images.

If you're reading this book in summer, you're walking down the street surrounded by people you don't know. You've put on your nice white trousers or your skirt or whatever you like. But you don't need any of these fabrics. It's hot. But hey, you're not on the beach so you find it natural to be dressed. In reality you cover your body with codes, with irrational signs. You wouldn't be comfortable without these signs, even if sometimes you find it a bit stifling under these fabrics. Being in shorts, or naked with a penile sheath covering your sex – there's nothing self-evident about it. Or rather, it's evident to each individual in their social body. Now, when the lady in shorts meets the Papuan dressed in his penis sheath for the first time, both find the other a bit surprising. A bit ridiculous certainly. Or frankly laughable. And it's true. They're both ridiculous since they don't have the codes. They don't understand the other's realities.

Each is a prisoner of themselves. Of their history. Of their necktie. Of their hat. Of their sheath. Of their bubble. Of their sphere.

The sphere is powerful.

Omnipresent, it splits our world into as many units all subject to their own self-evident ideas. To their dominant thoughts, both irrational and unconscious. We can probably never really extract ourselves from it. At best we distort the edges to touch the 'elsewhere', to see through it like a soap bubble. But the sphere can't pop like soap bubbles: it's always there. The sphere is what we are. Our matter, our mind, our gaze, our codes, our feelings, our reactions. It's the totality of our senses and perceptions. The totality of our interpretations and understandings. Everything that is outside the sphere is distorted and most often invisible to us.

The spheres of each society encompass all realities – not only common sense, which we easily understand as being exclusively made of constructions but even the physical senses. Our way of smelling odours, of conceiving them as sweet, violent, pleasant. Our way of seeing. Our way of feeling. Our way of perceiving time. Our way of being moved or not being moved by a butterfly wing. All this belongs to the sphere. We could say that an individual is a sphere floating in the great sphere. In the bubble of his or her society. You're starting to understand, I think.

Now what happens when two spheres meet? When two human societies that have never had the slightest contact meet? We can remember from the first contacts in the Americas that at first the two universes remain impervious to otherness, like two soap bubbles bouncing off each other. So at first they *don't* meet. They don't really see each other. They juxtapose without being able to perceive each other, without ever really conceiving each other. The first real interactions between two spheres are juxtapositions of meaning. Juxtapositions that don't communicate with each other but interpret, simplify, integrate, limit the Other to his or her own sphere of understanding of the world. If they do integrate it's like vinegar in oil and through the interpretative keys allowing each sphere to integrate the existence of another sphere whose divergence is not perceived for what it really is.

Men from the sky and white opossums

On 15 March 1493, Christopher Columbus returned to Spain and hastened to write a missive to King Ferdinand and Queen Isabella:

> They have never seen men wearing clothes, or ships like ours. [...] I took by force Indians from the first island, so that they could learn from us, and at the same time tell us what they knew of the affairs in these regions. This succeeded admirably because in a short time we understood them [...] and they did us great service. They now come with me, and they still believe that I come from the sky, despite all the time they have been and still remain with us. They were the first to say this wherever we went, shouting with a loud voice: 'Come, come, you will see men from the sky.'

Columbus is saying that these populations are not so different, and are reducible to the Spanish crown and to Christianity, in the same way that Caesar highlighted among the Gauls the presence of literate and philosophical priests and a god of commerce similar to that of the Romans, revealing civilizational and economic conditions likely to allow easy integration into Rome. The reduction of the other to oneself does not reveal an understanding of human unity here but probably reveals political aims that foreground a deliberately simplistic description of the peoples encountered. The 'you will see men from the sky' also allows Columbus to flatter his society and establish the crown's moral superiority. But beyond the political aims that may arise from such a presentation the phrase probably reveals certain much deeper realities. If the concept of 'men from the sky' clearly underlines the powerful misunderstanding of the local populations regarding this astonishing encounter, historical hindsight allows us to define, in a mirror, to what extent the Europeans too had no understanding of the reality of this singular event. It is the meeting of two universes that can neither theorize nor understand the precise reality they are confronted with.

If we identify a point in time, even an artificial one, in this first encounter between two distant cultural universes, the populations of the old and new worlds were separated by only about fifteen millennia. Almost the blink of an eye compared to the five hundred millennia that distinguished Neanderthals from *Sapiens*. Unfortunately, we do not have the precise words, the perceptions of the local populations. This conception of the 'men from the sky' was perhaps manipulated by Columbus in his project of conquering a new world, but it highlights how neither of these two populations was conceptually equipped to take the measure of the other.

If these fifteen millennia seem like a comma compared to the half-million years that separated, for a time, Neanderthals and *Sapiens*, these words remain valuable because we have only a few documents that allow us to imagine how two distant populations first set eyes on each other.

Unlike the Americas, the colonization of Australia doesn't provide us with a point in time that freezes, even symbolically or artificially, the moment of the first encounters. And here as elsewhere we never have access to the gazes from outside our societies – the gazes looking at us, telling us about ourselves. We don't have this feedback, this mirror that would speak about us, that would truly speak of the first contact. But this gaze, on all continents, is half-seen, glimpsed sometimes in the words of their witnesses, sometimes in the words of their children, those who today are only recognized between the lines as the owners of their ancestral lands. Columbus' words are a filter of his aims, his prejudices, his understandings, his voluntary and involuntary silences. But all words, all sentences, even mathematical ones, are filters. They are only filters to try to understand each other even if, as we sometimes understand, we can never understand each other completely. In the great land of Australia, let's listen to the contemporary words of Uncle Allen John Madden, a representative of the Gadigal, aborigines from the region of present-day Sydney.[5] Without wishing to offend him, let us reveal that Madden was born in 1949 and that he is imagining one of his group's first contacts in the 1790s, a century and a half before his birth. But the question is not whether we are faced with a first-hand witness, nor even whether what he reports was ever felt in this way by his ancestors who met Captain Arthur Kriddler Phillip in 1788 when the latter landed on the lands of his ancestors on the coasts of present-day New South Wales. We have seen that views are filters, and it's not a mathematical and well-ordered approach to history that interests us here. On the contrary, here it's the filter that carries meaning, that offers a gaze, as distorted as all gazes. Distorted also by the one hundred and fifty years separating Madden from Captain Phillip. But the words of 1788 were themselves filters. The structures of thought and understanding of the world of the Europeans of 1788 were not those of the Europeans of the present. And the filter of these words with a one hundred and fifty-year gap could well be preferable to the interpretation of the writings of a naval officer of His Majesty King George III. Of course, Madden's thought is a cumulative whole that is built on the very concrete tradition of his ancestors who did actually cross paths with Phillip. This cumulative whole is also built on the thousand stories that have been told since then in the cultural milieu of the aborigines of the Eora nation. Perhaps it's even an accumulation of many different stories heard here and there on the radio or online. Who knows? But it doesn't matter: these are words from the inside. Madden defines himself in what he calls 'the invisible people'. This time the invisible people are not those who have taken refuge in the distant mountains or the last remaining invincible forests. The invisible people are the ones of whom white people talk about in the past tense, in the streets of Sydney or elsewhere. Those who exist only in the negative, in the eyes of white people.

It still pisses me off when people talk about us as if we're extinct [...]. I was sitting at Alice Springs Airport with my brother and sister-in-law and overheard these people saying there are no blacks in Sydney. [...] It turns out they stay at Vaucluse and dine out at Watsons Bay, so of course they're not going to see any blackfellas. [...] We might be dressed different from those 'real' blackfellas that everyone seems to think only come from up north, with a spear and a kangaroo, but we're here. We've always been here.[6]

Is this how humans die?

Becoming invisible – as a result of their own will, their terrors, or, perhaps worse, due to the gaze of those people who do not see them. Who cannot see them, because they exist outside their sphere. Those who wouldn't be able to see aborigines unless they were riding kangaroos and shaking their spears at the sky. These are Madden's words, but this disturbing scenario is everywhere the same in its structure: whether we're in the United States, in Mongolia, in Ethiopia or in Australia, we pick up anecdotes that always tell us the same story. As if the scenario were always the same. The others, the different ones, the savages, those from the old days, all those strange and slightly dirty critters have disappeared. They disappeared a long time ago, with their feathers, their funny ways and all their ridiculous charms. Nothing remains of them. It's just folklore. Like those elves and leprechauns who no longer really populate Brittany and Ireland and who seem to have disappeared body and soul in the last dreams of wild children. Children of rivers and fields. Children from before the age of screens. The fairies died out with the last wild dreams of wild children. Thinking seems, here, to escape to lands that are quite unfamiliar but still lie at the heart of our subject. We have seen that when two spheres meet, each remains fundamentally invisible to the other. They don't really see each other. And if no one escapes his or her sphere, the sphere even seems capable of making what it cannot conceive invisible, of making it disappear. The perception of all reality is totally contingent, determined by unconscious but strictly constraining and delimiting structures. If societies believe in a divine being who resurrects the dead, walks on water or redeems the sins of all humanity, this being acquires his own – powerful – existence in the group. The reality of these supernatural, Christ-like or fairy-like entities is induced by the group's thought, which is transmitted essentially by perspiration, from generation to generation. This is how children's dreams, transmitted, strictly construct the reality of a view of the world. When Madden says that he was part of the invisible people, he's not using a figure of speech. He's not expressing a beautiful image full of poetry. He is putting his finger on one of the fundamental structures of reaction of human societies in the face of otherness, in the face of difference. He expresses – so powerful is the process – going so far as to make the different invisible – by which no one escapes the sphere. We are all in the sphere.

In the sphere, ironically, what doesn't exist can be the absolute reality and that which is divergent that fades, disappears, becomes invisible. Madden is one of the invisibles. There are no blacks in Sydney.

Let's go back to Madden's words. To his view of that first contact between aboriginals and Europeans. To that syncretic view weighing up a hundred and fifty years of pollution, of transmissions, but well within his sphere. In his gaze from the inside.

> The Blackfellas saw them coming up the coast and didn't know what they were. [...] Even when they got here, the Blackfellas thought they were possums going up and down the masts [...]. We had no idea if they were men or women. When you see a guy in a red coat, a hat and a wig, wearing a jerkin, you've got to be a bit crazy. [...] Governor Phillip made a good first impression with his gap-toothed smile. The initiation ritual here involved the extraction of teeth, and the first thing we notice about Phillip is that he's had this done. So he's a man of wisdom, an initiate. He's a man, so we ask him to drop his trousers, which he refuses to do, and asks one of the sailors to do it. [...] At that point, everything was fine. There was fresh blood in the neighbourhood, in modern parlance. But it didn't take long for the arguments to fester because we saw that these guys weren't going anywhere. They were here to stay. And it didn't take them long to clear all the trees from Farm Cove to the marsh where Hyde Park is.

Madden's words reveal a structure identical to Columbus's writings. On the one hand, we discern that the other is like us, and can be summed up as resembling us. But at the same time, something astonishing, something unspeakable, seems to characterize the other. Here the others were men from the sky. They were opossums. Fortunately, Captain Phillip had gapped teeth and so he was recognizably human. But he was still a frankly crazy creature with his red jacket and wig, and you needed to pull down his trousers to make sure that everything's in the right place, like with us.

Let's retain the elements relating to the structure of such events. The first stages of the encounter between two distant populations are uncertain. The two populations belong to such distant spheres that they agree, to a certain extent, to recognize each other as human, but as a little crazy, or as coming from the sky, or as engaged in activities that have no meaning and are only known among opossums.

What stands out from this first stage is that the spheres are water-tight. They observe each other in a game of distorting mirrors without understanding each other. Without recognizing each other fully, totally, as human. Or, if they *do* qualify as humans, this is actually based on a misunderstanding; Captain Phillip benefited, by happy coincidence, from gapped teeth, a human characteristic, indisputably, found only among human beings.

But the duration of the first stage, of the first contact, seems quite ephemeral and the incomprehension both in Columbus and in Phillip is followed by sometimes violent tensions. But this isn't always the case and contact can sometimes stop and freeze at this first stage.

This is the lesson of Ishi, whose sad epic we mentioned in the first part of this book. Faced with the Whites, the Yahi would lead a life of dissimulation, of lasting, perennial, total dissimulation, until the group was extinct. This is when peoples avoid each other, choose invisibility. Humans become *shadows*. There is then no second stage. The process of meeting humanities has no real continuation and stops here.

But when the first stage is followed by more in-depth contacts between populations, interactions will take place; however, in this second stage the two populations remain in their respective spheres, and all communication, all exchange, remains a game of distorting mirrors. Transmitters and receivers send each other codes that resemble long monologues where each side thinks that the other side thinks like them, thinks that the other side is like them. Even if they're 'a bit crazy'. This second phase, this game of monologues between humans could well be conceived as a series of misunderstandings. Transmitters and receivers involuntarily send each other signals that seem obvious to them but in reality have no meaning for the other. We observe each other without being able to see each other. We speak to each other without being able to hear each other. Essentially unexpected effects then emerge and can sometimes, over tiny details, tip entire societies into uncertain spaces. This is the idea that 'the gods must be crazy'.

The gods must be crazy

The Gods Must Be Crazy is a South African and Botswanan film released in 1980 that depicts how the discovery of an astonishing object soon tore apart an entire group of San hunter-gatherers from southern Africa. The intrusive object was a Coca-Cola bottle that fell from a plane. Brought back to the group, it's decided that the astonishing object was sent by the gods. A subject of curiosity and desire, the bottle is gradually involved in various daily activities. The director of this work of fiction, Jamie Uys, chose a Coca-Cola bottle as a representative, symbolic emblem of our societies and because 'it's a beautiful thing when you've never seen glass before'. The bottle serves as a rolling pin, a grinder and a musical object and soon becomes a source of problems, then of conflicts that danger-ously unbalance social relations within the group. Xi, the discoverer of the astonishing object, then decides to take it to the margins of the world to return it to the gods.

This work of fiction has been criticized as a white view of exotic populations cut off from the rest of the world. In the context of apartheid the approach was necessarily double-edged. The film also produces a vision of the noble savage whose harmony is broken by mere contact, however distant, with the Whites. In the 1980s, many of these societies were already falling apart and many San commu-nities found themselves dependent on food aid. Faced with colonized, marginalized societies, sometimes physically displaced to other terri-tories, the picture painted by the film did not denounce the sad situation of these populations. The anthropologist Toby Alice Volkman would say in 1985:

> There is, however, little to laugh about in Bushmanland: 1,000 demoralized, formerly independent gatherers, crammed into a sordid and TB-ridden homeland, getting by with cornmeal and sugar, drinking Johnny Walker or home-brewed beer, fighting each other and joining the South African army.

But this situation of traditional populations is a universal reality, one of those realities that we can integrate only with difficulty into the view that our societies share of the world. With forty years of hindsight, the situation of the San and a thousand other populations has hardly improved, a bit as if, in our unconscious, the place of otherness were just a tourist amusement. Dressing oneself in a few skins, adorning oneself with feathers or covering oneself with just a penis sheath, do not allow any integration into globalized societies. And folkloric forms, zombie and mercantile expressions of Western fantasies, are being maintained just as a guarantee for our moralities. In reality, morality has nothing to do with it. Otherness has become a tourist storefront. A financially profitable space summoning the shaman by exploiting the imaginary constructions of our societies. Experienced from the inside these cultural forms are only a vague echo of past realities. These societies are profoundly devastated. Experienced from the outside, in globalized international spaces, no real otherness is acceptable. The feathered ones serve only to liven up the post-ethnographic animation park. The feathered ones are turned into a spectacle and the tourist is entertained. Not only are these forms of otherness essentially moribund today but if the traditions of these societies really persisted they would neither be respected nor tolerated. I challenge you to try to get a plane decked out with nothing but a penis sheath – whatever departure airport you choose and whatever the destination. The airport has been sometimes defined as a 'non-place,' an artificial space where everyone pretends to comply with the most superficial illusions of our societies more than anywhere else. But the airport is above all a preamble. A project for societies that are finally disembodied, brought into line, simplified. An experimental zone where we can test the political possibilities for a future bleached of colour. Otherness can be given a home only within the limits of what is acceptable. In a nice television report, all clean and smoothed out, or artificially desiccated in the showcases of a museum, or to entertain tourists who are intellectually so limited that they think they can encounter a Savage just a few centuries or millennia down the line.

Do these words seem harsh?

In reality, they're still too polite. Too kind in the face of the realities of the planetary social ecosystem where otherness, if not reduced to an object that can be consumed, remains self-evidently abject and is instinctively relegated to the realm of exoticism. In a space that is by definition a bit dirty and disturbing. If the exotic escapes us, is free, fully exists, our unconscious forces it like a reflex act into the category of the 'not quite right', the rather ridiculous. We have always been, are still, and will probably remain forever, all of us, each other's opossum. The spheres never really understand each other. Fundamentally, they do not accept each other.

However, Jamie Uys based his movie on admiration for the 'Bushmen', the San, whom he encountered while filming animals in the Kalahari desert. The character of Xi is really played by a San Bushman, N!Xau and, if *The Gods Must Be Crazy* is an easy-to-attack romance, a certain encounter between the San and the Europeans did indeed take place, even if it tells a completely different story and Jamie Uys' words allow us to perceive certain conceptions that seem like a distant echo:

> 'At first [N!Xau] didn't understand, because they have no word for work', Mr. Uys said. 'Then the interpreter asked, "Would you like to come with us for some days?" [...] The airplane didn't impress him at all', Mr. Uys recalled. "He thinks we are magicians, so he believes we can do anything. Nothing impressed him.'[7]

The event and the words chosen send us back abruptly to the writings of Columbus, creating a half a millennium bridge between these two distant events: 'They now come with me, and they still believe that I come from the sky.' Beware of drawing any conclusion from the astonishing connection between these two events, which probably don't illustrate the way different aboriginal societies understand Western societies. These words come from within – from within us, from our sphere. They come from Uys and Columbus, five hundred years apart. They are filtered by our views, our conceptions, our selections. I'm not suggesting that these San and Taino did not put these words down but I am noting that these people must have said many things to Columbus and Uys. And that among the thousand things that were expressed those that Columbus and Uys retained are remarkably similar. The sentences chosen, selected, filtered, overlap remarkably and suggest not that the Taino or the San see Westerners as magicians or men from the sky but that it is Westerners through their mental filters who retain from a thousand sentences what, in the gaze of the other, allows them to position themselves as beings of a superior, ethereal, divine essence. As divine creations. But the selection of these two sentences is also a surprising echo of much deeper Western conceptions according to which humans were created by God in his image. Humans are us, always us. The injection of the biblical text does not in fact extend the divine nature to the plurality of humans. This myth of origins clothes *us* entirely, without projecting itself onto other humanities, whose nature must be defined, evaluated, positioned in contrast to *us*. The desire to project this divine nature onto other societies is hardly good news: it leads to the devastation of all cultural differences in a process of universal Christianization at an accelerated pace. The integration of others into certain concepts of our own societies represents a negation of the value of other traditions. This process of integration, this attempt to save their souls, to make them human, was and remains formidable throughout Africa and the Americas. And this right-thinking process, this voluntary destructuring of all otherness, is always set up in the name of the good. The good is obviously composed of universal values. Subjectively universal. And even today, unconsciously, structurally, it is God that we incarnate in our very flesh. A God who is not yet incarnated in the bodies of the aboriginal populations.

From Columbus to Uys, these two stories reveal us to be prisoners of our mental filters. We are unconsciously saying that the divinity of our nature is obvious in the eyes of the peoples we have encountered in our history. Strange glances that immutably echo through the centuries. We bear away in our magical, floating or flying vessels other humans who recognize as obvious the divinity of our nature. We come from the sky. We are magicians. These remarkable filters prevent us, block us from conceiving both others and ourselves simply for what they and we are.

It's funny.

It's sad, too.

Astonishing primates built in the image of God.

The gods must definitely be crazy.

Acculturation and the implosions of the spheres

It's a lovely story. But decidedly, otherness seems more than ever to be imperceptible, intangible, indescribable. Let's take another look, from a different point of view. This story of a single object that fell from the sky and was capable of disrupting all the social balances of a population – perhaps it could teach us something other than Jamie Uys' tragicomic nursery rhyme? Yes, I know: no one escapes the sphere and in the first stage of contact no real dialogue seems possible. The interaction of distant societies probably just expresses a game of uncertain monologues, a bit like in those tennis games in which we hit the ball to our opponent – who constantly just wants to hit it back. Except that in tennis, the players share the same codes. Spectators who don't share these codes and the game of social representations that assembles its audience fall asleep after the third serve. You have to know the codes. You need to have understood that as soon as you reach the professional level the stakes are not the ball. The stakes are the audience, the audience that watches and represents itself. The circulation of balls is just the excuse, the background noise of the social gathering. Isn't that so?

Let's go back to the meeting of two spheres that know nothing of their respective codes. Clearly, this astonishing confrontation of two distant societies can stop at this first stage, with one of the groups having understood that it understood nothing, or that there was nothing to understand and even less to gain and decides to become a *shadow*. This is the concept of Ishi and the Yahi.

But what about the second stage when the spheres, sometimes in spite of themselves, begin to influence each other without understanding each other, to alter each other, to aggregate bits of the other not in order to make it their own but to make something else of it? Something that makes sense in the neighbouring group's values and the way it understands the world. Could this second stage resemble what Jamie Uys wanted to express in *The Gods Must Be Crazy*? Ethnographic facts are always difficult to question from this perspective, mainly because the first contacts are very distant and so we can't analyse in real time the structure of events and their effects within human groups. In general, the most remote populations now surf the web to analyse *our* ways of constructing reality. We are also their subjects of interest and amusement. We have seen that the most remote, most isolated societies were probably so not because of their ignorance of the world but because they knew all about it and the group decided to distance itself from the global machine, to become *shadows*. But these societies are probably quite rare and are by definition indescribable. However, among those that are not shadows, many populations have been able to protect their own spheres and it is possible to examine the relationships they establish with their own objects and the objects of our societies. These relationships are not obvious and do not obey the cultural structures familiar to us. I was a sad witness to this discrepancy and it brings us back to that strange story of a bottle falling from the sky and destroying a population.

It's December 2005. We're leaving for the southwest of Ethiopia, in the basin of the Weyt'o, a river where several populations live, a mixture of sheep farmers, hunters, gatherers and warriors. I don't know much about this part of Africa but who can claim such knowledge? Ethiopia is an incredible gathering of cultures and languages. As you walk along its trails you see landscapes of astonishing beauty pass by. Well, you do so once you move away from the cities where the most unspeakable miseries are piled up.

A city, here, is a sinister window on the whole. A syncretism of all shipwrecks. Of all the slavery. Of all the stranded spheres, exploded into something that has no meaning. Something powerfully disturbing, something that can be wiped off the faces of others for a few dollars. But is it my gaze? In their words, in the bush, many will tell me they hope to get to the city. They have hope. It's like a piece of Europe within arm's reach. Two fathoms from Eldorado. The easy life, *la dolce vita* that you can see on the screens like an uncertain dream. Disconnected. With beautiful, clear, fresh water running from the tap, and all the rest. But in these cities of tin cans and concrete I saw only dismay. Haggard, dehumanized people. So, having left, they continue on their way. Maybe the uncertain dream is towards Europe. In fact, from a one-way ticket to *la dolce vita*, the path leads to every form of bodily enslavement. People's spirits are the first thing to get broken. It's as if many cities in Africa functioned as a hub of hope. Here they deal in cheap hope like a drug. People come for hope. So they take the hope of the people who flock from every region and imprison it in every cage of despair. There you are and there you'll stay. Human masses stretched out in tin huts and who have no other strength than to lie down and wait. Wait for what? They just wait. It seems that waiting is the whole purpose when you're broken. *La dolce vita* won't be in this boat. Nor in the next one. Nor in any of the others. It's Zola. But *Germinal* is a nice little story for oh-so-sensitive children. Here when we dare to look it's just unspeakable.

We're all guilty of these cities in Africa. Them. And us.

So we have to flee the city. And that's not why we came. Stanford University in California has recruited me to set up a research program in the Weyt'o basin. It's a total disconnect. Three spheres fundamentally sealed off from each other. These spheres are Ethiopia and Stanford and the third sphere is me.

Stanford has a little over seventeen thousand students spread across a huge campus of three hundred hectares. It's a very small university. The United States has about twenty million students. Less than a thousandth of American students are at Stanford, which is the size of a small French university. To give you an idea, this represents barely more than half the students at the University of Nice. The Stanford campus is an immense natural park strewn with centuries-old oaks and immense palm trees. Everything is perfect. I mean everything is simplified, clean. Even the insects seem to have been trained not to impinge on the scene. An almost disturbing impression of being too clean. Too ideal, too simplified to be pleasant. But there's no doubt it's magnificent. As you wander through these tamed artificial forests you come up against buildings that don't belong to any sensible definition of reality. Huge neo-Romanesque cloisters in cut stone. Huge church façades covered in gold mosaics. Eighty-one Nobel Prize winners have come out of this enterprise. Yahoo, Google, Hewlett-Packard, Netflix: all Stanford. The endowment of this small university is nearly 38 billion dollars. The entire budget for higher education in France amounts to less than 34 billion. We're here at the heart of a very special sphere. Fascinating. Almost frightening.

My first contact with Stanford was a large image printed on a T-shirt. A saturated blue sky, an immense façade covered with these gold mosaics, all lying at the end of immense avenues of century-old palm trees. First contact. First bit of evidence. It's a doctored image. But six months later I will be immersed in this offbeat universe. In this photoshopped image. Another surprise. No software had created this image. It was just a photo. A raw photo. Stanford is a projection in material terms, in stone and trees, of fantasized European imaginary constructions. A remarkable subject for study. I don't know whether Ethiopia or Stanford challenged me the most. Two fascinating spheres of humanity. I've just obtained my doctorate where I investigated the Neanderthals and I've become the observer of two remarkably divergent, watertight spheres that try to question each other, in vain. I find myself positioned as a go-between even though my sphere is fundamentally out of step with these two representations of the world. I paid my way through my thesis by playing the bagpipes in the dirty streets of Marseilles. Half street musician, half beggar. Always observant. You know, when you're part of the street, well, you see things. You're sometimes an object of amusement but most often invisible: you no longer think about your music. You observe surrounded by notes without really being observed. What you serve to passers-by, what they see, is an image, wearing a kilt. It's not you, it's not me. It's a cheap fantasy. As long as the coins drop in the hat. Sometimes a nice two-euro coin. A double sun. That's rare. Observing is what I do. Analysing. Deciphering the spectacle of everyday life with exoticism that challenges me as much as a Neanderthal. But to grasp it you have to move away. Move away from yourself. It's the same cleansing of oneself, of one's time, one's values, one's manners, of all one's mental constraints that I talked about in my previous book, *The Naked Neanderthal*. The little matter of all that revolves – of our sphere that covers us like oil. It's still there even when we think we are naked. Especially when we think we are naked.

And here are three unlikely spheres that are going to meet in the Horn of Africa. Two distant, uncertain universes and an out-of-place observer. We leave Addis Ababa, the capital, with our 4x4s, our drivers and our abundance of equipment. We carry every last drop of water in an incalculable number of plastic bottles swathed in huge nets. Theoretically, we won't eat or drink anything that's not sanitized. We assume that our bodies are not adapted to the local bacteria. Well … Personally, I have no idea. During my last mission in Africa, twelve years earlier in the Malian Sahel, the French team drank directly from a well without asking too many questions. A small tablet composed of chlorine and silver was dissolved in the flask to eradicate bacterial fauna. Nothing more. But I'm the only Frenchman in the team, and one just has to make do.

Ethiopia isn't really an adventure. The country is a conventional tourist destination and the 4x4s with drivers take more than eight hundred thousand customers each year on a neatly organized loop, ferrying them from one beautiful hotel to another, crossing superb landscapes. The disturbing visions of the city soon fade and marvellous human landscapes immerse us once again in colours and smiles.

Of course, the tour has the aftertaste of a walkabout. The route is conscientiously designed for the comfort of travellers. Tourists must be able to project themselves into a remake of Sydney Pollack's *Out of Africa*. A romantic Africa from centuries ago, embodied in the beauty of free, wild peoples and the aristocratic aloofness of the former Baroness Karen von Blixen-Finecke.[8] It has to be acknowledged that a century has passed and the encounter now has the taste of industrial crisps, the fat of mayonnaise on chips and the addictive spiciness of cold Coca Cola. These hundred years have dealt a nasty blow to all human societies. Not that it was any better before but from Africa to Australia via the Americas traditional societies have collapsed on themselves. As if the spheres for want of really being able to transform themselves had imploded. Poof. Just a dull thud and the feathered ones and their wild beauty disappear. I think that African cities are just that. They're the remains of the first peoples after implosion. After the universal Poof. They're the people lying in the streets waiting for nothing more than for everything to fall asleep in the dust of the hard ground. Their sphere has imploded and they've fallen to the ground. Poof. There they're remaining frozen, dazed, stranded.

African cities are the cages of the imploded spheres.

Plus, there are hardly any baronesses left who really pose as baronesses. Baronesses disconnected, in the distance, in the caress of their pearl necklaces and sable furs. Poof. As if it had all been just a dream. Digitized. Digitalized. Like digitalis, the foxglove. Foxglove is a dazzling flower. Foxglove is one of the most violent natural poisons. Like a warning sent from the floral to the digital. You see?

But along we drive, heading away from the cities, from their imploded spheres, from their elongated carcasses so stripped of humanity. And in each village there are the kids who run behind the vehicles. Hands outstretched. 'Heyyy!' they say. 'Hello! Helloooo!' Big smiles with all their teeth bared and their hands outstretched. The woman on my right who's come from California hears 'Hello!' in each village. I hear 'Dollar! Dollar! One dollar!' But it's true that they have a strong accent. And English isn't my mother tongue. She must be right. We're on the tourist trails. These paths across Africa are clearly marked, highly polished: they show what is presentable. It's true that these human landscapes are still disturbing. As we hurtle along in our vehicles, we see the clothes, the finery, the colours, the houses, changing at an incredible pace. I'm told there are about a hundred different languages in Ethiopia. And behind the languages, of course, it's all happening, an incredible human diversity is set in motion, on display, sung, spoken, laughed. They're magnificent, far from the cities. So, of course, I see they're pretty much glued to their television screens. That was almost twenty years ago.

Maybe these spear-wielding warriors now have smartphones stuck in the leather straps of their cartridge belts? I don't know. But this I *do* know: I see magnificent warriors parading with their huge iron spears resting on their shoulders. They walk along the tracks, as beautiful as if they were dancing, as beautiful as Jupiters with their lightning bolts pointing to the sky. For many of them the spear makes the man. The spears, their lengths, their shapes, change before my gaze depending on the clothes and the texture of the huts, the cabins, the everyday constructions. To each his spear. To each his language. To each his way of being a man. The cultural ferment varies from valley to valley. The Ethiopians certainly have as many spears as we have cheeses. But the beautiful weapon almost always comes in tandem with a powerful object. An object of power and desire. A magical object. The Kalashnikov. All the warriors with their spears, all the shepherds, all those who walk down these African tracks carry their beautiful AK-47s. Invented after the war, the Kalashnikov is one of the most widespread weapons in the world. An indestructible and easy-to-use weapon: a hundred million of them have been poured out on the planet. In Africa it's a must. An essential. It's the thing you can't leave home without. In 2005, the AK-47 was the smartphone of Ethiopian shepherds. They carry it nonchalantly on one shoulder, holding it by the end of the barrel as one would hold a stick with its bundle dangling from the end. They are superb, these ebony warriors with faces painted white and yellow, with their small metal barrels polished by the friction of never-ending daily use, by the continuous caresses they lavish on the wooden stock. They're repaired in a DIY style, sometimes with tape, often with fibres and ropes, or with hanging charms of hair, shells or feathers. They're not exactly a reassuring sight, but you wouldn't really call them weapons. They are clothes, sticks, adornments, signs. And, if the warriors' eyes sometimes have expressions that don't really invite us to return their gaze, their faces are often friendly, radiant with smiles. They're not on a war footing; the AK-47 is mainly a form of masculine coquetry like any other.

I happened to come across one of these warriors in the bush, alone, while I was looking for flints and prehistoric bones in eroded ravines. The nice thing about being alone is that the contact is radically different. There's a sense of togetherness: man to man, as male/female relationships aren't the same, they're complicated, there are codes, you see? And neither he nor I know the other's codes. We're men getting together – you might say we're both blokes but those are the words of my sphere. I have no idea what our meeting in the bush evokes for him. But we approach, we smile at each other discreetly, without insisting, without overdoing it. You shouldn't overdo it when you want a meeting to be a real meeting. We sit down, we share some tobacco. We smoke together, in silence. As if it were normal. As if we understood each other. Apart from the small plastic soles under his feet he's naked with an old rag placed in a ball on his head, perhaps because of the sun, perhaps to polish his Kalashnikov from time to time.

He's just dropped off a warthog, a wild boar from the savannahs killed a few moments ago. We have a long smoke. I think we respect each other, that a certain trust has been established. We finally exchange a few words. We mumble in English. I point to the Kalashnikov then the pig. No, no. He shows me his bow. Kalashnikovs are first and foremost ceremonial objects, messages, warnings; they are weapons only later. In general, our shepherds don't have the means to buy bullets. It was already a complicated business raising the funds for the rifle. This thing can fire six hundred bullets per minute. It costs a fortune. The AK-47 has to be seen. That's pretty much it. The rifle is embellished with a few pearl ornaments, the same that are on his bow, positioned at the same distance from the edges. The object is fully integrated into my friend's sphere but it's not really functional. Stuck along the butt, under one of those adornments, there's a bullet, just in case. It's very shiny, very polished too. It too, is adornment and warning. Like those multi-coloured frogs, too garish, that tell you not to touch them. Like the violent foxglove. Beautiful, fascinating, violent. No poison here. The poison is only affixed to arrowheads. But signs of respect. Signs that I respect myself and that I must be respected. In the final analysis, the Kalashnikov isn't a smartphone, it's probably a necktie – the sign of belonging to a group of values. A status. A way of positioning oneself. It marks a certain respect for certain codes, a bit like the fringe of fabric with which we often adorn our necks in our business attire. Of course, not everyone can access such an object of desire. We share the codes these objects represent in our respective spheres, we share its values of respect. Of virile respect. So as I walk along these tracks my eye begins to discern the real situation. It's not obvious at first glance but quite a few of these Kalashnikovs are made of wood. Entirely made of wood. They've been painted, copied, disguised, a metal barrel has been made by folding up the carcasses of tin cans. Don't smile. He'd like to be able to afford his own Kalashnikov and bask in the respectful gaze of others. To be seen as one of those who've made it. We all need that. Otherwise we wouldn't wear those ugly ties. Oh, I'm not attacking anyone, I'm not denouncing anything, I'm not pointing the finger at anyone. I'd hate to point the finger, even at a dominant thought. I try to look at the simple evidence. The lessons it teaches. We all need codes. Without them, there are no words. No exchange. No representation. The world without codes is that of these cities where the imploded lie. Outside the sphere, no codes or landmarks. Without a code, everything is replaced by the deal, by the minimal arrangement. And the minimum is the human being as the ultimate commodity. The death of the old traditional codes seems to be able to leave room only for easy trafficking where everything is freely exchanged without constraint. In slavery and the denial of all hope.

Over the last century, the poof has been total, universal. Its breath has spared no one. It has left the human people haggard. All peoples.

But in some corners, a hope, or a period of waiting before the death agony, something perhaps has survived. Is it an echo? Perhaps a shadow?

Men in the trees

Our archaeological surveys will allow us to move away from the tracks on which tourists go round Ethiopia in circles. As soon as we move away from the main tracks, the balance changes. Tour operators don't stand out for their originality. People don't change formulas that work well. And here we are faced with human geographies that are unsuitable for any improvisation. Even if the area is touristy, Africa can't be improvised.

We won't find any BnBs or nice little hotels off the beaten track, where there's only one track provided for this purpose. Diverging from the main route is neither planned nor possible. And, in just a few minutes, the distance from the circular walkway allows us to encounter an 'elsewhere' that we might have imagined was extinct.

Could the lost worlds still be alive? Just a few steps away? Right there, on the side of the road? I've always been intrigued by otherness. I've tried to find it, of course. In every corner. In the 1990s I tried to escape to every accessible corner. In the Sahel or in the Gobi. Towards Siberia. People darted questioning glances at me there. I hoped to see the remains of sunken worlds but all I encountered was sunkenness, without any of the remains of the ancient epics. That's not quite true: by pushing the limits, exploring, advancing further into the corners, I did encounter things – remains of ancient times, of ancient gods, still lay here and there, out of breath. Rarely where I expected them. The confrontation with otherness sometimes befalls you at a street corner, in the most unexpected places – one evening during the 1998 football World Cup, in Brittany, or in an old Irish valley now hidden, stifled, between endless rows of prefabricated houses. And it's not necessarily in encounters with the last shamans of Mongolia that the confrontation with otherness finally finds its full intensity. The gaze has a lot to do with it … No, it's much more than that. From now on, otherness is above all an education of the gaze. Its re-enchantment. Its capacity to look sideways. To read through words. And it's this sideways gaze that allows us to look at ourselves, straight in the eye – allows us to discern, in our society, the elements that structure its own exoticism. So, of course, the experience leads us to extract ourselves from ourselves, from our values, from our conceptions. From our sphere. We don't come back unscathed.

Maybe we don't even come back at all.

But this time, otherness wasn't just around the corner, but really around the corner of the bush.

Moving away from the tourist trails, we reached the Weyt'o basin. The analysis of satellite images in Stanford had allowed us to identify areas of erosion hidden in the bush. Areas where the run-off of water had allowed the soil to be cleared and dug naturally to reveal its archaeo-logical treasures. Our archaeological explorations targeted these natural ravines to identify their contents. Easier said than done. The trails have an end. Then it's the bush. Stifling. We deploy our machetes, crawling along, crossing rivers. We have to define the areas where there shouldn't be any crocodiles. But we can't always see to the bottom of the river. And our Maale guides themselves don't always seem sure of their move. Yes, yes – it's an archaeology really similar to that of Indiana Jones. I'm the first to be surprised, but in the final analysis the series of films isn't just folklore, even if a cross has never, ever, marked the location of a treasure. On the other hand, our satellite images were right on target. The erosions were there and indeed contained some ancient and particularly interesting archaeological furniture. But to reach our crosses and cross a few kilometres of bush we had to reckon with hours of hiking under a blazing sun. It's a tale worth telling. Arriving at one of these erosions, I see incredibly fresh flint objects emerging from the ground. Bifaces, stone tools and all the remains abandoned by distant hunters from the Palaeolithic. Marvellous remains of prehistoric camps probably more than three millennia old, spilling out into the bush. Reaching this area had required a forced march of more than five hours under a burning sun. And we had to leave it to get back to our camp before nightfall. It was obvious that we wouldn't be able to return to this deposit that we had named Igaro VI. A few photos would not be enough to establish the initial documentation of such a discovery. We were exclusively prospecting and didn't have authorization to collect archaeological furniture. I suggested to the team that they head back, but leave me here so that I could draw the most characteristic elements freehand. I would join them. I knew the direction and I could walk faster than the team.

There's a bit of teeth-gnashing, but no one's going to make me leave now. The matter's settled. I sit down on a trunk and start my drawings. The immediate problem is the mosquitoes that cover your body as soon as you stop moving. They're kept away by highly toxic military oils. But they're still a problem. While the mosquitoes remain relatively easy to cope with, being alone in the bush brings about other realities. About thirty minutes after the group leaves, the sound of footsteps and broken branches becomes audible. They get closer and closer, in concentric circles.

Baboons. Baboons are among the most aggressive of all our cousins. And the noise of the branches was not due to the heaviness of their movements. They were warning me of my intrusion into their territory.

Still, they're nice guys, baboons. They warn you before shooting. They size me up. They approach. They start to chuck branches and a few stones to the right and to the left. But they're well equipped! They've chucked a biface, damn it! I have to speed up my drawings, and I regularly utter a few cries in my hoarsest voice. We warn each other, on both sides, of the limits not to be crossed. But you have to come to terms with the residents – and they're undoubtably strong. Fortunately they still doubt their own strength a little, just enough time for me to document and bear witness to a few fine archaeological pieces. We've come to terms. The intruder is me. And what's more, they're the strongest. In hindsight, it's a delightful tale, and it allows me to depict the living environment of the populations among whom we have settled for a few weeks.

This Weyt'o basin is occupied by several cultural groups. We are among the Maale, naked warrior hunters with bows. My first encounter with them quite amazed me. My geographical wanderings, from the Gobi to the Sahel, from the Irish west to the Anatolian highlands, had confronted me solely with the remnants of old traditional populations. With the shadows of their shadows. With the agonies of their past glories – when anything original still remained. In this little corner of Ethiopia I was also brought face to face. My perceptions of a desperate globalization had to be revised.

We've been off the tourist trails for some time and are approaching the river Weyt'o. An old tree can be clearly seen from a distance, at the top of a rocky massif that our 4x4 is lumbering up. Soon we'll be skirting its mass of vegetation. But I find myself projected into another place. My throat suddenly tightens. Speechless. Hanging on the wide horizontal branches of the old tree are about ten warriors. Standing. Naked. Quivers of arrows across their backs. Bows in hand. Naked warriors dressed only in bows and arrows. The long-awaited *elsewhere*. The elsewhere that I've stopped expecting now looms over me with the greatest naturalness. Wide smiles inform us of a warm welcome. Around the tree are objects, their objects. All made of wood, skins and fibres. Really? So it really exists? Still? Naked warriors living in sync with their hunts and the power of their bows. A sudden hope of freedom. Of otherness. The writings of Malaurie, Paul-Émile Victor and Lévi-Strauss, despite the sadness of his tropics had cradled my entire life journey, imbuing me with elsewheres populated by humans who did not merge with, or limit themselves to, our conceptions. Encounters with extinct othernesses. A thousand geographical wanderings had neither brought me to nor prepared me for these fragile flavours.

In the coming weeks I would be able to take advantage of this proximity to soak up their ancestral knowledge. Not that the West isn't ancestral. We're all ancestral. But they were finally, different. Another path. Another story. Their manufacturing processes for these bows, the Kossi, and these arrows, the Ichi. I count nine categories, all adapted to the hunting categories, from the Pumka with a wooden plate at its tip to stun birds, to the Sissiagwi, veritable twisted needles at the tips of the arrows to pierce small prey, lizards and snakes. There are their poisons, their firewood, the Surung where archers rub their hands to obtain the first embers to light the Tami. Fire. Moments of discovery. Moments of intense sharing. But make no mistake. They are not free from any contact with the Whites. The missionaries came long before me. Probably in the 1930s. A long time ago in any case, and so here they are – good Christians. There must probably already be Bibles in Maale. And no memory of the ancient gods. By dint of insisting, I get some vague notions from their collective memories. The old beliefs are, as it were, passed under the table. With a sort of embarrassment. No, no, they insist before telling me about the creation of the universe, it wasn't really a religion. It wasn't much. Strange stories. That's all I'm getting.

A few kilometres away the neighbouring people are no less interesting. But the territory of the Tsamaï adjoins the great tourist track, impacting more directly on their old traditional structures. They're beautiful, the Tsamaï. Men and women with their bodies covered in pearls. Pearl warriors. Warriors with spears.

Regularly, on the pretext of lost or stolen sheep, they go to war. But war is also a social necessity. It allows for peace and the exchange of women. The genetic survival of groups and especially the transmission of traditions. And the Tsamaï are regularly devastated by floods of Maale arrows raining down from afar on the spear-wielding warriors who need physical contact with the adversary. I'm told that the last skirmish in the 1990s left a few hundred dead on that hill over there. The bodies were abandoned on the hill. It has since become taboo to approach it. Relations seem to have become peaceful again. But, to take a wife, a Maale must bring back the testicles of a young Tsamaï, I'm told. And vice

versa. A remarkable way to keep the expansion of the younger generations in balance. Old ways that would now be forbidden. But you still
keep your long dagger close to you, held by a fibre around your waist.
Just in case. An opportunity may arise from time to time. I don't know.
Those are their words. During one of our archaeological surveys, I see a
kid a few years old covered in white and blue beads. A Tsamaï! On Maale
lands! That's quite something. My Maale guides immediately correct me.

'He's a Maale.'
'But he's covered in Tsamaï ornaments!?'
'Yes, but we're at peace now.'

This information was remarkable. Ornaments are considered by many
prehistorians to be markers of the deep identity of the social group.
Markers that allow us prehistorians to distinguish distant cultural,
even linguistic, groups. A possible gateway to an inaccessible, past
ethnography through the analysis of crafts, tools and weapons whose
technologies would have less value in the recognition of societies
carrying the same cultural baggage. These people told me a completely
different story. Finery is shared. They adorn themselves with the finery
of enemies when they become friends. But never, ever, will a warrior
with a bow become a warrior with a spear. The peace sealed by two
distinct cultural and linguistic groups has its own limits. Weapons define
men and also a certain notion of humanity. But now you'll remember
my stories about Kalashnikovs. And you know that weapons are also
ornaments, above all ornaments. They identify the group as much as the
place of each individual in his or her society. When we identify in our
distant Palaeolithic societies the adornment of the rest of the material
productions of the social sphere, we may be fumbling around, asking
the wrong questions. Weapons and beads and all the rest. Each gesture.
Each anecdote is part of the framework building a society. Defining a
humanity. They're each partial components belonging to the systems of
representation of individuals and societies. None of these components
can be distinguished, extracted from the matrix in which it makes sense.
Especially when times are peaceful.

But might we sometimes escape the sphere? See the other spheres?
Take inspiration from them, swap ideas, transform ourselves? I don't

think so – maybe like chameleons, superficial mimes, like in those masked balls or in those children's games which we played, imagining ourselves as Indians, cowboys or distant metallic warriors sailing towards Alpha Centauri. Besides, this Maale covered in Tsamaï pearls was a child. A kid before puberty. Not yet a man. Would he still wear these white and blue pearls after having been welcomed into the circle of the grown-ups? After the initiation that would make him a man – after he had brought back some Tsamaï testicles?

We have reached the limits of our concepts in which armaments and adornments, societies and linguistics, wars and peace define societies and some of their potential for influencing each other, transforming each other. We're at the gates of the transition. You know, those transitional industries that might or might not be Neanderthal and that would mark the transformation of this humanity into Ourselves before gently dying out. The transition. Acculturation. These are uncertain notions. Poorly defined in the face of what a human society is. And what can we say when we project it not onto two human societies but onto two humanities separated by five hundred millennia of divergence? How will they influence each other? Will they actually influence each other? Will they even be able to look at each other? I don't think so. No one escapes the sphere.

The humanities don't really look at each other. Don't really understand each other. Maybe. Because my words always dance with doubt. They open paths that everyone is free to explore.

The gods must be crazy 2.0

This story has a sad ending. Not only that of the Neanderthals or that of the people of Weyt'o whose fate I do not know. But that of my too short stay among the Maale. We had seen our flints. I had tried to look at the magnificent peripheries of these populations. But we had to leave. Our 4x4s were topped with mountains of empty plastic bottles threaded into vast nets hung on the roofs of the vehicles. Our waste. But our waste had made people envious and negotiations had been initiated without my being informed. Several Maale had come to ask for bottles. Their magnificent wooden receptacles were no match for these receptacles, a manna as translucent as water and as light as air. No one had said a word

173

to me about it until the day before our departure, but the question had been heard. You'll get our bottles. All our bottles in those enormous nets – but after our departure, to avoid the daily demands, the jealousies. And the immense net now fed an immense desire. It was the object of all the stakes, of all the strategies developed to approach it. *The Gods Must be Crazy* 2.0. For real. Jamie Uys had thought up his bottle of Coca Cola in a sort of African prophecy. A Maale guard has been placed for a few days in front of the nets. Faced with so much wealth, promised and soon to be accessible, a riot was brewing to get hold of the remarkable objects of their desires. It seems that the team is happy. It's a good bargain. We're returning without bulky waste and honouring them with our generous donations.

It's the end of the wood. It is the end of beauty. And everyone is happy. So there is so much thoughtlessness. We're all academics. All thoughtless. It's not even cynicism. A well-meant bit of thoughtlessness. All this artisanal beauty, this timelessness, or almost, that I had sought so intently was going be wrecked because of this. Because of us. Because of me?

What a bunch of idiots.

But don't they see anything behind their big kind smiles? We're leaving. We're happy. Are humans simply kind? Kind and blind and thoughtless?

I was going to have to act alone. And above all, not be seen. They wouldn't have understood. Neither my American colleagues, nor those naked warriors, the bow hunters who had coveted our plastic objects for weeks. I had to wait until the middle of the night. Wait for the protection of the night. Arm myself with my knife, in the shadows. And silently slit the throats of this enormous pile of plastic bottles, one after the other. Slit the throats of our nauseating projects. Alone.

I'm not bragging about it. I'd never spoken about it. I remember it with nothing but shame. Did I do the right thing? Who am I to impose myself like this, in the shadows, both on the naked warriors and on my American colleagues? What happened when our vehicles crossed the horizon?

In July 1951, Malaurie had told the general in chief directing the strategic American base of Thule located in the heart of the Inuit territories to 'go home'.

Another time. Another place. Another death agony.

Seventy years have passed. No one had gone home except the Inuit. To the upright houses of the Whites. You get the picture.

My last stand, my 'go home', was no less desperate.

But here I had not come alone, like Malaurie or like Victor or like Lévi-Strauss. 'Hell is other people', said Sartre. It may be other people. But hell is we ourselves, too. Our doubts. Our wanderings. Our belated actions. Our limited awareness.

I had drunk the bottles too. What happened? Afterwards?

Do the Maale surf the internet? Are they friends on the web with the Tsamaï? What a crazy idea of mine it had been to try and freeze this moment, this society in another place. An *elsewhere* against my despair in the face of a global reality whose ugliness I feel like a slap, like an absolute constraint insinuating itself everywhere. Like a total project. Totally totalitarian.

Did I act for the Tsamaï? My *Tristes Éthiopiques*.[9]

I like to think of myself performing these actions, of course. Taking this role. But I must wash myself of my pretentions. I must accept responsibility for their darkest side also. An elsewhere for me. For my own pleasure. Finally. An elsewhere that I constrain one night by the blade of my knife. I'm the idiot. It's oneself. Always.

Wanting to preserve is already appropriating. My role in history isn't really much more enviable. Carrying the weight of my ego. But I knew full well that by going to meet the Other – we change that Other, and we change ourselves.

In our disordered memories[10]

But these visions, these ways of preserving or appropriating the world here or in boxes, in our museums of stone, are deeply rooted in our Western traditions. It's our history. Our relationships with ancient relics. Our desires to tame the world, to catalogue it. To simplify it. To freeze, identify, record. To preserve? That's the luminous face of our simplifications, of our appropriations. It's all us. Again.

In Europe as in North America, immense collections of objects from all the Amerindian nations lie accumulated in remarkable museums. Millions of objects from the thousand cultures of the first nations. Those from whom the land was stolen. Those from whom memory was filched.

I've been hanging around in our museums for a long time, you can imagine. Archaeological collections. Ethnographic collections. Tying time together. Putting it all in boxes. Everywhere. I've hung around the reserves of American institutions for a long time, between thousand-year-old flints and wooden and leather objects from the last century. You're often alone there, but cosily surrounded by the thickness of the history of many worlds. From time to time they pass by, the Native American Indians. They come to see. They come to see if they can remember. Their footsteps echo on the metal floors. I usually hear only their footsteps. And above all, their silences. They say nothing. It's astonishing, by the way. I look, sideways; their eyes are wide open. They didn't know. They didn't know that there was so much here. That so much had been gathered from their old memory. From their old story. From their old life. Their old life – well, what they imagine of it. They don't know. They don't know anymore. They look. They see. And something sings in their hearts. And something weeps in their hearts. An echo, a myth. An invention, of course. It doesn't matter. They are their mothers' mothers. Everything we think is true, in the end, when we really think it, with sincerity. What matters to them now, apart from the dream? You see?

But the Whites think they know. And the objects are there in any case. There. In these iron corridors. Leathers, barks, skins, woods, paintings, earthenware. In iron cabinets. They come to seek a memory that is no more than a shadow. No more than a whisper. A whisper that grabs them by the guts. Like when you never really knew your father. Like when you never really knew your mother. They look lost, with a lump in their throat when they advance in these metal corridors. And me with them, with all my ridiculous protective coverings, over my hair, on my hands, around my shoes, my white coat. In my disguise of a guy doing research, all covered in white and blue. Am I not ridiculous, covered in plastic at every extremity of my body so I'll be allowed to handle objects from the olden days? You have to play the game. Play at pretending. Pretend that it's all useful. Pretend to believe that it's not just some kind of slightly ridiculous rite performed by those white people. They, too, are inventing a memory for themselves. A memory that they have also lost. A memory that they, too, invent. They too have not really known their father. Not really known their mother. They too are uprooted. Amnesiac, like these Native American Indians dragging their feet along these iron corridors. Far from their land. And they know, somewhere, that all this is shameful.

I don't like this. I don't like what we do. I don't like what the people in the museums do. With an air of kindness they talk and look … How can one put it? With a kind of discreet condescension. Kindness. The worst. Like when you talk to children, when you don't know how to talk to children.

There's this 'repatriation' program. Giving back to the Native American Indians the objects of their nations. Little by little, of course. Reluctantly. I have heard some of the things they say, that show neither emotion nor intelligence. 'It's ridiculous, it's folklore, it's a re-creation, it's an invention, they mix everything up, they no longer have any tradition, they don't even know their languages, they invent Indian ways that only exist through what they've seen of them in Hollywood. It's all a load of bullshit …'

It's true that, when I take a discreet look, I see that they are wearing jeans, or a Stetson, or a few turquoise necklaces. They may have bought these turquoises at the museum's souvenir shop. From this museum or another. It's true that they speak English, I can hear them. It's true that they may no longer know what these objects are. They may no longer know the crafts. They may have lost their intelligence gleaned from their fathers. The sweetness of their mother's words. It's true. It makes my throat tighten but it's horribly true.

But when I look at the woman guiding these Indians through these corridors … What can I say …? What memories does she have of her roots? She clearly has Italian origins. But what has she kept? What memory of her fathers and mothers? The painting of the Mona Lisa in 1503, in Italy, coincides with the first contacts between Europeans and peoples of the Americas. The charming person who guides our Indians doesn't really look like Mona Lisa, I must say.

She's … how can I put it, disguised in pink candy. Pink glitter shoes, a pink sweater and jacket over a pink skirt. Pink lipstick is the last angular touch to the painting. It's clear that, while her knowledge of the traditional populations of the Americas is very superficial, she can't really claim the cultural heritage of Renaissance Italy either. She's probably an expression of the exact opposite. She's a pleasant enough person and we talk from time to time. Obviously, she's a pure American product, but she calls herself Italian. She thinks of herself as Italian. Deeply Italian, sincerely Italian. She even knows how to say *Buongiorno*, which is quite something, in itself, when you're American.

Do you see? Do you see it now?

We are all Indians without land and without history. Without the memory of our mothers. Without the memory of our fathers. We are all those sad clowns. Abandoned by all memory. We are all those guides dressed in pink candy and these lost Indians in jeans and Stetson.

We are, all of us, that. With our disoriented landmarks. With our disordered memories.

Could humanity be summed up in this creature constrained by two great two profound destinies? An ephemeral creature, living only a few years – and an amnesiac creature not knowing or being able to preserve its own memory?

The times of extinction

Yes, I know. I know the sadness of these observations. It's painful when we question our possible trajectories. But this book is a sad book, you know that. Even if I try to laugh about it. Sometimes. You can't come up against the greatest extinction of humanity without looking yourself in the eye a little when you get there. If you do get there. And this book oscillates between laughter and tears. Real tears. Those that come up against human darkness. In our instincts. In our actions. In our history. In our cultures. But these words are drawers. We politely place them in their wooden boxes. I told you. It's our thing to classify, to categorize. To simplify the world in order to understand it a little. To make ourselves believe that we understand it a little.

Culture is a dirty word. I mean an indefinite word, a box, our box. Our simplification. And everything that obeys the sphere. Culture is neither language nor just the uses we make of it. Culture is perspiration. The perspiration of its ancestors, of those thousands of generations leading to us, transmitted in the things we say with our conscious mind but even more in the unconscious unsaid.

We are the genealogical matter that flows through us without us being able to see it or define it too precisely. And our ways of being are perhaps only the surface part of much larger realities totally beyond us.

178

Millennial, unconscious transmissions that overflow onto the individual. The death of the Neanderthals performs a break in these millennial unconscious chains. Of these chains uniting us to our pasts. But not just that. The death of Neanderthals does not exist. Not in so many words. The imaginary scenario of a *Sapiens*/Neanderthal face-off induces us to understand these humanities in an essentialized way. Two realities. Two watertight drawers. Two blocks. But what if Neanderthals were not a humanity but, in themselves, a plurality of humanities that slowly distinguished themselves over tens of thousands of years? A differentiation that was not just cultural. A differentiation of biology. Of bodies. And what was the place of the mixed-race people? And that of the mixed-race offspring of mixed-race people? All of them humanities that are essentially invisible.

Of course, with the Neanderthals, cultures die. Societies die. Jared Diamond has pondered these collapses and their theorizations. His words speak of societies. Of large and small civilizations. From Rome to the little traditions of the north. These deaths, these extinctions, are mere commas: the collapse of societies, of organizations, of certain ways of being human. But Neanderthals were not a civilization. Neanderthals were *hundreds* of civilizations collapsing like dominoes.

And it was in their ways of being and in their flesh that they collapsed and died out. The collapse of societies and of their cultures probably represents the easy part of the story. The visible tip of the iceberg. The visible part – even if it can be terrible, too. Mere collapses, the mere deaths of human societies. But the collapse of societies can follow its own logic to the bitter end and extend, sometimes, to the extinction of bodies. 'Ishi and his mother remained together until her death [...]. Then he found himself alone, without a companion, for the rest of his hidden life.' This hidden life would last for nearly three years. And here, as for the Neanderthals, the collapse of societies goes hand in hand with physical death. The death of bodies. In seclusion from the other world, ours, a world of which Ishi would never feel part. 'He felt so different, so distinct from everyone, that he refused to conceive that he was like the others; it didn't seem right for others to think so. "I'm someone, you're the others, it's in the inevitable nature of things" – that's more or less the way he judged himself.' The collection of these thoughts by Théodora Kroeber suggests that Ishi already perceived in his own way the irreducible concept of the sphere. As if the history of his people represents just one of the countless anecdotes through which we can observe that, beyond societies it is sometimes bodies, those of all men and those of all women, that die. They die too.

But why do humans die? What can be lying in the unconscious background – that force, so powerful, so deep, leading to the eradication of all otherness? Always this terrible process in which the man-eaters cut things up. We are all man-eaters. But what is this power of eradication that we perhaps carry within us? Is it possible to verbalize it? And, if we can't understand it completely, perhaps we will learn, just perhaps, to be wary of it. If we can find the words to define this dark part of Ourselves. Think of all the words of Lucien Bodard in which he describes, speechless, human beings in their raw reality as man-eaters. Why all this?

> These Indians remained more or less free from their masters, until around 1940. Because that is when the last genocide began, the one that is still going on. The cause of these massacres? Not an Eldorado, not just the notion of wealth and greed. But the modern world, mere progress. All these complex, contradictory, all-powerful forces that constitute the march toward the West.

And Bodard lays out the events that take place hidden under the immense forest canopy. There's the explosion of physical bodies. The implosion of all social bodies. But what is this march toward the West? What is this modern world that cannot fail to happen, inevitably? What is this progress that seems to work like an invincible steamroller, crushing and swallowing human flesh. A human machine. A jammed machine that just can't seem to stop. The Native Indians die because of human machines. Mental constructions. Our logic. Our imprecations to advance into the forest. We don't even have the excuse of El Dorado. We kill to kill. We destroy difference. We eat the spheres for want of being able to encompass them. Couldn't we start to perceive some broad outlines? A blueprint of replacements? Simple and, at the same time, infinitely complex, since these processes of collapse have never been explored except from the outside. With an emphasis on the most visual, the most obvious, the easiest factors. Pandemics, wars, acculturations, famines, maladjustments to the new world, to the mutations that advance and, under the forest canopy and everywhere else, eat human flesh and human souls. And the thousand excuses. The thousand reasons. Could they ultimately be just the effects of something deeper? These circumstances when understood as causes all seem increasingly terrible. All these natural impacts. Terrible. But harmless if we investigate the deep causes of past collapses.

If we are to believe many of the discussions, Neanderthals themselves evaporated, freeing up space for climates that were collapsing on themselves. For swept-away biotopes, for pandemics, for volcanic explosions, for stellar radiation. And it was all those things going *bang!* that on this view led to the eradication of all human populations. But not ours. And all those processes, all those biotic anecdotes that include our humanity as one of the factors, among many others, of the greatest extinction of humanity. So did Ishi die of cold, too? Of hunger? Yes. Oh yes, obviously, bodies die from all that. So yes indeed, Ishi and all his people died just as all the humanities still hiding under the forest canopy die. From those effects. Ultimately, this is how humans die. Having run their race, they had to die, all their bodies had to die from those effects.

From those effects.

So we're looking at things from too close up. Can we really investigate the causes?

T. S. Eliot suggests to us that 'this is the way the world ends,/ Not with a bang but a whimper.'[11]

The suggestion is totally counter-intuitive. If we transfer these ideas to the Neanderthals, and other fossil humanities, we should not look for any great volcanic explosion or any other loud noise coming from outside human societies. We shouldn't blame the collapse of the Earth's magnetic shield or the thousand climatic pulsations of our past. Neither this humanity nor our humanity will die out in its famines. Nor in an immense flash of light sweeping over all our human constructions. These effects, as terrible as they are, cannot represent the deepest causes of the great processes of extinction. They do not confront the deep logic of these exterminations. The terrible Hiroshima was not an end, but a beginning. Nearly eighty years after the terrible flash of light Hiroshima has become the main industrial and shipping hub of Japan. Neither thermonuclear implosions, nor the deepest conflicts, nor powerful volcanic explosions, nor the tsunamis, nor the great famines of Kangi, Kanshō, Tenmei and Tenpō led to the collapse of Japan. And the small island has remained a planetary power. Strategic, economic, cultural. Could the power of their mental frameworks, the power of their sphere, make this civilization immutable? Inaccessible to any ultimate collapse? Embedded in its powerful cultural constructions, the sphere of the rising sun lets the most terrible crises flow across its empire. The same crises that struck our distant civilizations and from which we can outline their collapses.

And what if the poet was right? Let's follow Eliot. Civilizations do not disappear in loud bangs. From Rome to Greece, from Egypt to Mesopotamia. Slowly, without really making any noise, they collapse in on themselves.

This would be the case in culture as in nature. However, we're familiar with these biotope collapses, these encounters with invasive species, and all the processes leading to all the extinctions. No less than 99.99% of living organisms that have lived on planet Earth are extinct today. And if we often remember this great meteoric bang that eradicated the dinosaurs in a dazzling sidereal flash, in general they die slowly. And in silence. In their deafening murmurs.

The collapses of human societies have been widely theorized from Arnold Toynbee to Jared Diamond;[12] they view collapse as a result of the accumulation of environmental factors and in the human interactions that result from them. But Diamond like Toynbee note that, behind all these processes, all these anecdotes, more profoundly, 'civilizations die of suicide, not of assassination'. In an inability to meet the challenges of their times. An inability for which they seem to be solely responsible.

But here we must ask ourselves: could this notion of collapse not be itself a construction? A category too hastily accepted in our ways of thinking about time, of thinking about societies and their millennial successions? Could we not be blurring these realities in the very conception of what a collapse would be?

Let's step back and take another diagonal look. The nineteenth century.

Whatever has become of the nineteenth century, just a tiny temporal cable length from 2023? The nineteenth century is right there. If you were born in the 1970s, all the old people of your childhood were these children of the nineteenth century. You rubbed shoulders with them. You saw them die out.

And, despite this proximity, despite the physical presence of the nineteenth century, our political regimes are profoundly different. Our technologies belong to a completely different universe. Our arts and architectures are a thousand miles from the Impressionist and academic schools of art. It's not even certain that an ordinary person these days would really understand the words of an ordinary person from the nineteenth century. If we were to consider the most grandiose technological and cultural expressions of the nineteenth century, we might well wonder with some concern about what has become of this remarkable civilization, mysteriously extinguished without it being possible to trace its extinction or understand its evaporation.

So, how do human societies tip over? How do political systems collapse? Great architectures? Linguistic structures? Arts? Technologies? Religions? The poet warned us. In a whimper.

Perhaps, in reality, all this is experienced, has only ever been experienced, in whimpers. These whimpers could almost feed the illusion that civilizations never really die. Do not really know an end. The great linguist Georges Dumézil noted the incredible persistence of very ancient spoken traditions, whether we look to the Caucasus or to a thousand other regions of old Eurasia. Under the dominant ideas, forms of Zoroastrianism still live in Iran. The divine Ahura Mazda still orders the mental universe of tens of thousands of people there and the potentials of Zarathustra are still feared by the Ayatollahs.

Everything dies and everything persists. Always. In powerful whimpers.

Could it be possible that we find ourselves, faced with these collapses, confronting our own intellectual constructions? In our need to catalogue things and define a beginning and an end? To point out limits? Could our visions, our expectations of great collapses not be nourished by our ways of conceiving the reality of the world? Of apocalypses and divine acts of vengeance sweeping away humans each time they have sinned? Could the mass of collapses also be our way of looking when we look back at the temporal immensities of our past? Temporal thicknesses that overwhelm us and in which we try to cut out pieces of cake to better understand them and more easily digest them?

Are we, here too, even here, prisoners of the sphere? Wrapped up, framed in the thickness of our mental constructions that lead the gaze, always, in the same directions?

The concept of collapse only very partially tells us about the old history of our humanity. Collapse doesn't tell us how human societies ultimately are connected together and reproduced. We see them, however, these connections, these tipping points in the history of humanity. From the great Neanderthal extinction to the tipping point of the universe of hunter-gatherers of the Palaeolithic into sedentarization, into agriculture, into livestock breeding. Goodbye Palaeolithic, welcome Neolithic. Much more than collapses, these two events of the upheaval of all human societies confront us with the most sensitive connections in the history of humanity. The prehistorian Jacques Cauvin described this other shift. That of neolithization and the abandonment of all the old ways of being in the world. And he noted this: *In the beginning was the myth*. Rejecting environmental constraints, food pressures, climate change, Cauvin no longer places societies on the margins of processes that they simply undergo as the natural fruit of the long history of humanity; he repositions our mental constructions, our thoughts, our immaterial structures, at the heart of one of the greatest social and economic revolutions in our history. At the beginning was the induced myth that the shift of human societies from nomad to sedentary, from hunter to breeder, obeyed immaterial, religious, mythological necessities and whose outcome produced one of the most profound, social, technological and economic reshapings of human societies.

And it is this prevalence of thought, of myth, of the understanding of the world that would lead our humanities to explore and shift into radically different universes. As one human being. Astonishing?

Could collapse, reshaping, capsizing be processes subject above all to our own views? Would the capsizing of civilizations ultimately be above all an internal collapse of our conceptions, our beliefs, our myths? Above all an implosion of our spheres – a collapse of the mental spheres of a society? This collapse of all the imaginary structures that construct a view of the world and unite populations together to form a society? This process whereby the sphere implodes is expressed by shame at oneself, the development of this sense of the infantilization of traditional populations, well documented in ethnography, which marks the withdrawal of a society in the face of new ways of being in the world. These processes of collapse and domination do not only affect the materiality of populations, their crafts, their knowledge, their ways of being. Mental dominations, dominant thoughts, are at the heart of the processes generating the forms of infantilization and shame among dominated populations.

These processes can themselves be supported by the mythological conceptions of the populations that would mark time. The Mayans saw their gods and their predictions in the advance of the Conquistadors. Their collapse was fostered by their own mental constructions. We see here the meeting of imaginary processes and those carried by ideology. Carried by beliefs, in the certainty of the coming of an inevitable destiny, however illusory. The disappearance and replacement of oneself, as a matter of course. A sense of history. It's the 'march of modernity', as Lucien Bodard put it. No one can go against the march of history. But this march is irrational, unconscious. Shared and accepted as a matter of course. This obvious tendency that guides us towards a destiny that we feel to be inevitable is actually constructed by our imaginations alone. But shared imaginations. Common. Powerful. Bulldozers in our unconscious. And we move forward – worse, we obey these mental constructions all together as a single human being.

A shared vision that makes everything else invisible. That's how the sphere works.

And yet there, on the sidelines, a large part of the Amerindian nations have neither died out nor vanished. They're still there and perhaps it isn't their forced confinement in reserves provided for this purpose that generates their greatest erasure. Could it not be, as obvious as the nose in the middle of one's face, their total disappearance from all our imaginations that induces the erasure of all these nations? The erasure of the imagination of the Whites but also their erasure in their own imaginations. As if their erasure were the natural fruit of the march of history. Who determines this march? No one. We all create it together. In our spheres, those structures that connect us and frame us without us even being able to see them. Without wanting to. Unconsciously. It is beyond us and seems to advance on its own, outside of us like a powerful, irresistible mechanism of self-destruction sweeping away all the mental constructions that made a society possible.

We're a long way from the bangs and the bams and the bings.

Can you see it now? Can you see the determining factor in the capsizing of humanity?

All our myths, all our beliefs, all our conceptions are hidden here. From the most material to the existence of fairies and other creatures constructed within us, from Iceland to old Ireland. If we investigate the disappearance of the little people, who perished with those who carried their memory, we are plunged into mental processes that are in every way comparable. The little people, and all our other constructions, our beliefs, our religions, invite us to explore our imaginations and the imagination of all our disappearances. The disappearance of knowledge, traditions, religions, of what once forged our certainties about the realities of the world. Those immaterial constructions have nevertheless structured all our materialities, curbing lives, raising menhirs, pyramids and cathedrals in the most obvious materiality. Are these stone giants not the fruits, in our landscapes, in our materiality, of our wildest dreams? Are these senseless, dazzling, overwhelming, profoundly useless construc-tions not, above all, the fruit of our imaginations alone? It is these imaginary constructions that frame more than ever our perceptions of all reality. Might not the collapse of all these imaginaries, all these myths, all these beliefs, all these conceptions of the world be the invisible force that in the first instance quietly bears all human societies towards their own collapses?

Societies would not collapse in any material contingency, in any shock, in any war, in any famine, in any epidemic.

All these great bangs, these poofs, these bams, these howls, this horrible yammering, these noisy death agonies might just be the countless terrible anecdotes of the history of human societies. All these great bangs are always there in our daily lives. They are our terrible anecdotes. Dreadful. Anecdotes, stories that have never forged the destiny of any society.

We were warned. Human societies collapse, in Eliot's words, in a mere whimper.

It's in the implosion of their sphere, this invisible framework that governs each human being, each shared way within a human group of conceiving the world that humans seem to die. The imploded spheres are those millions of people lying haggard on the ground of African cities and all the peoples lying down, without will, without vision, without hope, lying down in alcohol and drugs. From the aborigines to the Inuit, from the San to the Indians of the great forest. And all those stranded in the 334 reservations of the United States. In the misery of the territories of San Carlos or Pine Ridge, of Tohono O'odham or Standing Rock. Here daily life is no longer punctuated by anything but expectations. Without a goal or a horizon. It is the implosion of their spheres, of their traditional framework of conceiving the world, of all their conscious and unconscious mental structures, of their cultural frameworks. Their spheres have collapsed in on themselves. It is these implosions of the sphere much more than all the famines and all the climate changes that lead people to their inevitable suicide.

In the end there are neither great bangs nor great collapses. It is in the collapse of their views on the reality of the world that humans die. Would it be enough for humans to see, to glimpse, other humans whose understandings, whose spheres are so radically different from their own that their own visions of the reality of the world are dramatically affected? And capsize, inevitably, towards somewhere else? Capsize into what is now understood as the new meaning of history. The meaning in which everyone will be engaged. Irresistibly. As one human being.

And here it is our certainties and our realities that are tipping over. What if until now we had understood merely the anecdotes affecting the history of all humanities? The surface – that surface, terrible, nauseatingly chilling but which, like Hiroshima, never impacts on the depths of our ways of being in the world? What if the essential were elsewhere – in the mental spheres that guide us and carry us away in spite of ourselves?

The implications of such thoughts are troubling. Of course, we are creatures of flesh and blood and need earthly nourishment for the survival of our bodies. But, by our nature, our social constructions surpass us, always carry us at a forced pace towards an elsewhere of which we are not aware.

It's as if our mental constructions alone surpassed all the other contingencies of our world.

Will setting these words down, laying bare the unconscious structures that seem to frame us, to direct us, allow us to free ourselves from them, just a little?

EPILOGUE

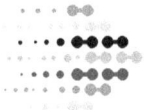

This is how humans die

We have been travelling down some unexplored paths. From the spheres that encapsulate us to the mythological constructions that establish our legitimacy and that, at the same time, build up our forces of eradication.

I warned you. It's a sad book.

We are living in zones of turbulence. A deep turbulence, affecting our conceptions of the reality of the world and of human nature. This line of thought also deciphers what prevents us, most likely, from constructing precise, objective representations of the world around us. It lays bare the impossibility of being able to look at the world free from our powerful mental filters. Our invisible filters, which direct us, organize us. In spite of ourselves.

Setting our social constructions at the heart of the processes affecting human societies, we must then conceive of our humanity, in its long history, as largely emancipated from natural, climatic, biological, environmental contingencies. We cannot conceive of humans outside of the mental spheres that encapsulate them. Spheres which connect us and which, at the same time, make us blind to any form of otherness. We can then see humanity, our humanity, as lying in the power of its thought. But this thought is all-invasive and it drives us towards normalization, towards standardization. It pushes us into line. In spite of ourselves. It's a thought from which we cannot so easily escape.

Here we are then in our own mental cages. Cages in which we are carried along, all together, as one. This is a property that maybe gives our humanity a remarkable efficiency. An incredible power. Perhaps here is the force that in the shadows allowed us to take precedence over all other forms of past humanity. A force that would also perhaps be our greatest fragility.

Thus, *Sapiens* societies could well depend and have for a very long time on their mental constructions alone. Functioning as entities profoundly emancipated from their environments. Our conceptions of the collapses of civilization always modelled on the history of human societies and the contingencies of their natural environment could well have failed to see the deep causes of these processes. After all the cataclysms, all the pandemics, all the wars, all the famines, the collapse of human societies would depend much more deeply on the implosion of their mental spheres. Spheres that govern, organize, all the conceptions from which human societies are structured, unconsciously.

But would Neanderthals for their part have been more fragile, more dependent on the changes in their natural environment?

If we follow this line of thought, *Sapiens* societies don't adapt to environments: they impose themselves on them. These are degrees of course; current and past human societies can adapt, obviously. But in *Sapiens*, social constructions, mental constructions, impose themselves on the environment. And the comparison of these mental structures with what emerges from Neanderthal crafts, their powerful dialectic with rocks, with materials, raises the question of their own way of conceiving the world. Did Neanderthal impose, as we do in spite of ourselves, their imagination on the material world? Or did they adapt to it?

If we follow my thinking and my way of understanding these Neanderthal crafts, such populations did not impose their projects on material reality frontally, by the power of their minds. They compromised. They adapted. They created. Or rather, they brought out the mind in matter. They transformed their creativity to the realities of the world. Bringing out of each material what they perceived. As if they were following, step by step, in a remarkable dialectic with materiality, their own Rorschach images. Perhaps these were their most profound arts. Their forms of creativity.

Could such thinking give a certain weight to naturalist theories, those based on landscape and environmental factors, regarding the extinction of this humanity?

According to common conceptions, collapses and extinctions should only be based on historical processes necessarily distinct in each society and in each region of the world. These historical structures could then have led, theoretically, to the emergence of very different conclusions.

But one can only imagine what those regions of old America would have looked like if the local populations had largely overcome their European settlers. But, despite the vastness of the Americas and the diversity of the nations that were spread across them, this was not the case. It was almost as if those historical chains were inevitable and signalled a direction of history that no one could hold back. However, this direction of history could not have existed outside of our own conceptions. Of our own mental constructions. More than the succession of this or that historical event, would it not be the power of those mental constructions, and their acceptance, in the end, by all the local populations, that induced a certain direction to the history that began on these two continents? The power of mental constructions, the power of our myths that ended up annexing all forms of divergent thought, sweeping away – as if all this were natural, as if there existed a meaning of history – all those populations and their old conceptions of the world. Somewhat as if the mental constructions of humans could spread in each of us, against us, against ourselves.

Like a viral expansion of our mental constructions. A pandemic of dominant thoughts.

And this brings us straight back to the history of the twentieth century. To the power of the ideologies that from communism to Nazism imposed their rhythm on humanities. Ideologies: thoughts that nothing seemed able to stop. Leaving humans alone against the irreducible direction of history. And yet, all of this, these immense structures, these immense totalitarian scaffoldings, belonged only to our mental constructions, projected onto matter. Configuring the materiality of the world with our minds alone. Like the menhirs, the pyramids, the cathedrals.

No one can go against the sphere. No one can go against the meaning of history, the viral expansion of all our mental spheres.

Let's go back to the distant Palaeolithic lands. While the old *Sapiens* encountered an incredible diversity of human societies across Eurasia, the conclusion of this process was nevertheless universal. The extinction of all cultures and all humanities. Could this collapse also have been based on the confrontation of humanities? On the shipwreck of mental constructions, ideological constructions that structured the unconscious universe of local populations? A shipwreck of the myth. An implosion of the sphere. And it is probably the ethological divergences, the profound divergences between these humanities, which could be the reason, projecting Neanderthals, in their game of distorting mirrors with *Sapiens*, into their own impasse.

Did they perceive, unconsciously, that they were confronted with another version of themselves? Not a more human version but a more efficient version.

The processes of normalization of *Sapiens* behaviour that I highlight confer on our humanity, in fact, objective advantages in all technological and logistical areas. We must at the same time note the dangers that come with this property of unity, with its standardization of conceptions and ways of being in the world. A danger that is also expressed in our own societies, in a certain relationship to natural spaces. In a certain relationship to our biosphere from which, as Descola has noted, it is madness for us to separate. Not only is our relationship to the natural world around us dangerous, but also our human matters, if we are not careful, can be even more dangerous.

In these games of distorting mirrors, in this balancing of *Sapiens* conceptions and Neanderthal conceptions, we can recognize and update elements that seem to structure our own natures. The *Sapiens* ways of being human. But what does the highlighting of mental processes specific to these two humanities lead to? Can we ask how Neanderthals died out and what they took with them?

The Neanderthal extinction cannot possibly have corresponded to a singular event. This extinction was necessarily the result of a succession of events, historical events, potentially expressed over a relatively short period of time. It is chilling to note our inability to define precisely any single one of these events with assurance. We conceive of the archaeologists as beings who seek a history, we imagine them seeking the origin of human things. With Neanderthals we are faced with an inversion of this meaning. It is indeed the end, the end of all things, that we seek and this end does not in any way announce the beginning of something else. Our humanity unfolds its existence in complete autonomy and is not born from the extinction of the other creature. It's like a paradox. Parallel stories.

In 2014, I contributed to a vast synthesis published by the journal *Nature*, with forty-six other authors covering the main sites of the whole of continental Europe. It concluded that from Russia to Spain no traces of the old Neanderthal traditions could be found after the fortieth millennium. The study was remarkable for the redefinition of the temporality of this extinction. But this formidable tool could not account for the processes at play in this twofold extinction: the extinction of human traditions and the extinction of humanity. In the present work I have analysed this process through the discovery of the remains of a Neanderthal body. The discovery allows us to confront the limits of all our analytical methods, and to follow the construction of our questions with regard to some major discoveries. The discovery of a body; the discovery of remarkable genetic divergences – this confrontation with an unexpected plurality of humanity among the last Neanderthals raises profound questions. How could Neanderthal populations occupying very close territories have ignored their neighbours for more than sixty millennia? Something here raises questions not about *Sapiens* but about the structures of Neanderthal populations. Their ways of being in the world. Their isolations. And at the end of their course, as the only conclusion of these long processes of societal and genetic differentiations, their replacement by *Sapiens*. A replacement that we calibrate at the Mandrin cave not on long human generations, but as a temporal snapshot.

It's incredible to think that the replacement of one humanity by the next could well be summed up in a few seasons. The speed of the process sweeps away any hypothesis addressing the replacements of forms of humanity in climatic or environmental terms. It isn't slow afflictions that ultimately lead to the extinction of populations.

We also discover from the material in this book that, in Europe, the last Neanderthals represented a plural humanity. Several deeply divergent Neanderthal branches occupied the continent at the time of the establishment of these processes of extinction. Populations that were plural not only in their cultures, but in their flesh. The discovery of these sixty millennia of differentiation within populations living in neighbouring regions is dizzying. So we could see Mediterranean Neanderthal societies heading up the Rhône towards continental Europe but without these movements leading the populations involved to cross paths to exchange as if Neanderthal societies no longer communicated, as if they were rooted, among themselves, in territories without communication. This incredible distinction of populations had profound implications for

Neanderthal social structures and their relationship to the territories in which they seem to have isolated themselves and taken root. Do we not have here a fundamental cause allowing us to think about the Neanderthal extinction?

But if we follow my thinking around those astonishing creatures, the Neanderthals, we have to conclude that the archaeological clues from which we include them in our way of being in the world could well be only the result of our projections. Of our constructions. These projections would directly affect our representations of the last Neanderthals and their connection to the culture of the Châtelperronian, one of the first so-called modern cultures. The analysis shows in fact that the Châtelperronian presents very clear affinities with traditions of the Mediterranean Levant that we know are connected to *Sapiens*. If we follow this line of thinking we have to conclude that the Châtelperronian corresponds to a migratory phase of *Sapiens* in Western Europe, thereby excluding, in fact, their connection to Neanderthal populations. This goes against current data from the hard sciences, genomics, proteins and dating, which today form the basis for the idea that these traditions are linked to the Neanderthals. My analysis is therefore posed as a prediction. But prediction is a fundamental property of science. This ability to predict may even be the best definition of the scientific approach. Will the hard sciences be forced, as with the understanding of Thorin, to mark time again, as against the human sciences? Such comparative analyses had already allowed me to predict the connection of the Neronian to the *Sapiens* populations several years before the analysis of the human remains of the Mandrin cave. History will decide.

But the Châtelperronian is only an entertaining anecdote. We may have to envisage that all the so-called transitional industries could mark the establishment of the first *Sapiens* populations across Europe. According to this proposition, Neanderthal societies would then have died out without showing the slightest mutation of their ancestral cultural traditions. The historical model built up over decades that postulated a remarkable mutation of Neanderthal societies before their extinction should then be completely abandoned. It would therefore be necessary, in the footsteps of this work, to completely reconsider the structure and meaning of this extinction of humanity.

But the fragilities of the Neanderthals represent just one element of the equation. Rejecting the naturalistic theories, the climates, volcanoes and epidemics that could have impacted on this extinction process, also leads to repositioning *Sapiens* at the heart of these questions. The human societies that crossed paths across the vastness of Eurasia were, nevertheless, biologically and culturally very distinct. These social and cultural diversities affected both *Sapiens*, the bearers of many traditions and the fossil societies at the time of their extinction. Those humanities that would soon die out were the heirs to many traditions. To clearly distinct traditions. And yet, beyond the Neanderthals, it was all the cultures and all the ancient humanities of Eurasia that would die out. Neanderthals, Denisovas, Flores and all the others. All the humanities that were parallel to us. It should not be possible for such a diversity of humanity, such a diversity of cultures and civilizations, to be swept aside so simply.

And yet, they died.

They died out, all of them. None of the fossil humanities would survive. Here too the convergence with the *Sapiens* expansions cannot be bypassed. The encounter between these humanities must be placed back at the centre of the equation. The weight of this encounter in the midst of the evaporation of all otherness cannot be prudishly avoided or naively marginalized as just one of the many factors of a very complicated equation. Focusing on climates, epidemics, volcanic explosions and all the anecdotes impacting the history of human societies amounts to casting a prudish veil over our own history.

At the heart of those extinctions of humanity, those systematic, total and rapid extinctions, the *Sapiens* factor was probably central.

Could something fundamental in *Sapiens* have escaped us and still need to be brought to light? Something that far exceeds that species' technological capacities and knowledge alone? It is not enough to note that *Sapiens* is the key element in all extinctions. We must understand how, whatever the environments, whatever the biological humanities concerned, whatever the cultures of those populations, *Sapiens* universally took precedence over all other humanities. We must explore what this revolution in the fiftieth millennium is telling us. The diversity of environments, biological humanities and cultures affected by this process implies, in fact, that the answer to this astonishing equation does not arise from the geography, climates, or technological and social traditions of the humanities that will be swept away. It is in the phases of this remarkable expansion of the *Sapiens* populations that we can document the extinction of all other humanities across Eurasia. As if something in this process of expansion had systematically worked to the advantage of the *Sapiens* populations.

Could all other humanities have been more fragile than our humanity?

In this book I have established that the *Sapiens* of the Neronian mastered the oldest archery in Eurasia, also noting that they settled with weapons and baggage in Neanderthal territory, populations that used only heavy spears wielded directly by hand. You might have thought I would come clumping in with my big boots, with my martial technologies distinguishing fossil humanities from modern humans. But here too we must reverse our concepts. Explore the depths.

I am not suggesting that *Sapiens* were technologically superior to the now fossil aboriginal societies that they encountered across Eurasia. My thought is elsewhere and it perhaps lies at the opposite of such simplified linear visions. I never build my thoughts around overly simple equations. The bow and arrows are in no way the technology that led to the superiority of humans over the Neanderthal creatures. This is where we must scratch a little deeper below the surface and look at the divergences that seem to be very deeply rooted in the processes of standardization that are visible only among modern populations. Could the emergence of the remarkable technologies mastered by those old *Sapiens* result from this way of conceiving the world? From these processes of standardization that find a remarkable echo in the technological productions of our own societies?

Are these technologies, therefore, just the visible part of the iceberg? The easy part to watch and understand – even if the demonstration of such technological divergences required more than twenty years of analysis and the construction of three doctorates around the technologies of these societies, their positions in time, and the structure of their weapons?

The proposal already made in *The Naked Neanderthal* never pointed to any possible technological inferiority of Neanderthal populations over *Sapiens* populations. Neanderthals showed a remarkable mastery of their crafts. The divergences that I highlight between these two humanities do not relate solely to the technological history of these societies. We are confronted with structural divergences rooted in the nature of those humanities. Divergences that tell us about the ways of conceiving the reality of the world of human populations. The standardization that is expressed in a remarkable way in *Sapiens'* technologies obviously does not tell us about their technological knowledge alone, but about the mental structure, the ethologies of our own populations. My gaze does not focus on the tip of the arrow nor on what it could have allowed our distant ancestors to do. My gaze questions the process itself. The profound process that would induce the extinction of any creature that was not *Sapiens* across all the territories of the vast expanses of Eurasia.

Since I exclude any natural factor from this process, only *Sapiens* remains in the equation. But to say that *Sapiens* supplanted all humanity, any creatures that were not *Sapiens*, is a bit too simple.

We must ask more questions. We must go to the bottom of the abyss and see crudely what is happening there, stare it right in the eyes. Stir the entrails even in their nauseating odours. Understand what it means, without sugaring the pill – this proposition that the expansion of *Sapiens* directly generated, across the planet, all the extinctions of humanity.

In the history of *Sapiens*, the proximity of technologically asymmetrical societies has almost systematically led to the supplanting of the most fragile societies. Here we have the entire history of colonizations, of the two Americas, of Australia. This observation has only rare exceptions. It's not that the bad guys get rid of the good guys with the help of fearsome technologies, nor that the strong eat the weak. But humans regroup themselves, instinctively, in their conceptions of the world, based on efficiency. The efficiency that ensures the survival of human societies. Their reproduction. Could human societies confronted with more efficient versions of themselves therefore lower their guard, instinctively, from the inside, leading to the collapse, in spite of themselves, of all their cultural frameworks, their values, their myths, their constructions? Could an enemy from within push them, unconsciously, to abandon what they are, so as to make way, inevitably, for the most efficient versions of our human organizations? Here the processes of reproduction, of standardization, which seem to structure our humanity could well dictate the future of our civilizations when they meet. We would be faced with behavioural structures, ethologies, specific to our humanity, which guide it, in spite of itself. Ways specific to *Sapiens*, ways that push societies to die out of their own accord in spite of themselves in the face of social constructions perceived as more efficient. We instinctively side with the winners. Of course, the perception of these superiorities is a construction, a myth. And even more, an unconscious edifice. But at the moment when this construction, this invention of our perceptions, in spite of ourselves, finds itself shared by the dominant and the dominated, the sphere that turns out to be weaker implodes of its own accord. Could we be touching the invisible meaning of history? The meaning of history, the one against which no one can do anything, against which all are helpless? The meaning of history, that machine that crushes everything before it and that nothing seems to be able to stop? That machine that seems to create, in the background, invisibly, the whole framework of our destinies in spite of ourselves? The machine that pushes us in the shadows towards other inventions, other mental constructions, other spheres?

Something here distinguishes *Sapiens* not only from Neanderthals but perhaps also from all the fossil humanities that *Sapiens* replaced on the planet. A behavioural specificity of *Sapiens* that makes it different from all other humanities. This way of being in the world, this way of conceiving the reality of the world, this way of refusing all divergence, represents a simplification, a unification of planetary realities. This way of standardizing ourselves, of refusing all difference, does not seem discernible in any other humanity. We probably have, here, a behavioural specificity proper to our *Sapiens* populations. A particularity. An anomaly compared to other humanities. An anomaly that induces our need to reproduce the dominant behaviours expressed within the group. Boris Cyrulnik explores, in equally chilling texts, these almost unspeakable components of human nature.

This view, this proposition is hardly enchanting. They speak to us very crudely not of our societies but of our human nature and of our future.

When asymmetrical relations are established, let's look at the history of all our colonizations where the fate of the most fragile peoples seems irremediably sealed. Could our own colonizations represent a stuttering, muted form of those old Palaeolithic stories of human expansions across Eurasia?

In our recent history, in our colonizations, such asymmetries are, however, based on fairly superficial divergences. Cultural divergences, technological divergences. In our old history, in our 'prehistoric history', these processes were not expressed only in the technological and social structures of human populations. They affected different biological humanities. In the old Palaeolithic story, these asymmetrical relationships must then also raise the question of the ways of being in the world of our humanity in relation to all other humanities.

But *Sapiens'* crafts, techniques and standardizations would then only be the visible part of much deeper realities, revealing structures powerfully anchored in our humanity. Our social balances, our exchanges, our music, our dances, our ways of thinking are deeply governed by these ways of being in the world. Exploitation, systematization, the systemic exploitation of all the resources of an environment, optimization, would not be specific to current societies but could well be part of the structural behavioural properties of our humanity. In our flesh. The proposition is counterintuitive, as the organization of human societies can seem to vary with each cultural tradition. But cultural divergences do not suggest the non-existence of a human ethology. They reveal the possibility, through our social constructs, of softening or exacerbating behaviours deeply rooted in us. Culture versus Nature. And this possibility alone expresses here a glimmer, a hope in humanity.

My thought brings bad tidings about the very nature of our humanity and it will perhaps be fought against, or pilloried. I will be criticized for having said things in this way. 'It is impossible but that offences will come: but woe unto him, through whom they come!' This portrait of our humanity probably comes as a shock and, moreover, finds little echo in the classical definitions of human nature.

But my words do not aim to point to some original guilt in our humanity. They are meant simply to sound a warning. Neither pandemics, nor environmental changes, nor volcanic explosions, nor our sterility, nor the shifts in our climates will ever be able to put an end to the reign of *Sapiens*. Our descendants will inherit our defects and the altered geographies that we will have created. But humans, like the distant Neanderthal creatures, will probably never become extinct in the terrible circumstances of the moment.

The simplification of the world induced by the extinction of all other human creatures now leaves us exclusively facing ourselves. And, yesterday as today, this crossing of millennia suggests that only humans can make humans disappear.

Sapiens against *Sapiens*.

Our need to reproduce dominant acts, to standardize ourselves, to act as a whole group, all together, has been able to give our humanity its remarkable efficiency. Its deepest strength. It probably also constitutes our dark side. Our Achilles heel. Our profoundly totalitarian potentialities.

I cannot mention the extinctions of humanity without a pang in the heart. But this pain does not belong to the gaze fixed on these extinct creatures. This pang in the heart stems not from what I understand of the dead but from what I feel for the living. For the sole survivors of those distant extinctions. A pang in the heart, too, at the nature of *Sapiens*. A nature that confronts us with the darkest dystopias. And with all the horrors yet to come. I warned you. This book is a sad book.

But I am the father of two beautiful children and these words refuse to look only towards our darkest horizons.

Is there only despair in the face of our deep nature?

In the face of our ethologies, of our need to reproduce the actions of each individual, in the face of the herd and the mass movements carrying each one away, as a single person, there remains, perhaps, a hope. It's the hope of the black sheep. Humans have experienced many realities. They have invented many societies. They have elicited so many ways of being in the world. Thousands of different human realities present themselves before us. They are spread out in our history and on all continents. What these thousands of societies were proves that our Natures can be subjected to our cultural constructions. To our mental spheres. These words are not meant to be a banal reminder of the way humans are each other's frenemies, but as a note of hope.

In *Sapiens*, does not the will to reproduce acts, to do like everyone else, lead to the possibility of a profound overthrow of our societies, of our ways of being, acting as one? A possible implosion of our own sphere – an escape?

So yes, our vital need to form a group, to be one, all together, confronts us with the worst of what lurks in us humans. But sometimes, sometimes, when our cultures bite our natures, could this vital need not also carry us, as one, towards the best?

Will the black sheep be able to carry the whole flock with it?
Hope, perhaps.
Because this is how humans die.
But this is also how they look towards tomorrow.

Acknowledgements

These writings evoke human beings, in their destiny and in the limits of that destiny, but also in their hopes for freedom. On this path they encounter memories and surprises and sometimes, of course, they lose their way. On this perilous journey, I have to thank all those who, from one crossroad to the next, have guided me, inspired me, pulled me out of a thousand ruts. Those who built me up. Grandmother Yvette and her iron character – but a gentle iron character. Grandfather Maurice, made of kindness alone. We always miss them, our elders. They live in us forever, too. My friend Valéry who became a brother and whose paths are always crossing mine, at a distance. To all those who tossed me a coin when I was bagpiping for my supper. Each of those coins offered me a little freedom to think, write, and sketch out the distant past. The Rosen clan is dear to me, too, as I follow my path. This book explores unknown lands. It opens doors to the unexpected and allows us to rethink what our humanity looks like, in various ways. My sincere thanks to Zac Pelleriti for his final insight in transcribing these words into the language of Shakespeare, in an effort to preserve the precision and poetry of my words originally conceived in the language of Molière.

Finally, may the Maale forgive me ... for the bottles.

Notes

First words

1 This is a quotation from a poem by Louis Aragon (1897–1982) set to music and recorded in 1961 by Léo Ferré (1916–1993): the refrain of the poem goes 'Is this how men live and their kisses follow them from afar?' (Translator's note.)

1 Unexpected complications on the way to the unthought

1 Every year, archaeologists provide their administration with a report on their archaeological operations. In 2006, I wrote: 'The discovery suggests the possibility of microchronological research with a hitherto unprecedented precision.' It would take us eleven years to publish the first results and thirteen years for Ségolène Vandevelde to draw all the implications in a thesis defended in November 2019 at the University of Paris-Sorbonne. See Ségolène Vandevelde, 'Y'a pas de suie sans feu! Étude microchronologique des concrétions fuligineuses. Étude de cas: le site paléolithique de la grotte Mandrin (France)', PhD dissertation, Panthéon-Sorbonne-Paris-I, 2019.

2 The Mandrin cave is in the French *département* of the Drôme, in the Auverge–Rhône–Alpes region of southeast France. (Translator's note.)

3 CNRS: the Centre national de la recherche scientifique (National Centre for Scientific Research), the biggest state organization in France of its kind. (Translator's note.)

4 This is what I wrote in 2017: 'The lines of descent that can be proposed here are no longer of the order of the generic idea of productions (producing predetermined pointed stones from technological systems still partially rooted in the old Levallois modes) but illustrate a strict replication of systems; the technological systems of the Neronian in the western Mediterranean are similar to those documented at the beginning of what is recognized under the name of the Initial Upper Palaeolithic in the eastern Mediterranean. [...] It would then be possible to envisage the existence of historical patterns in which the arrival of the first *Homo sapiens* societies in Europe did not occur at the very moment of the Neanderthal extinction, but preceded this extinction by seven to ten thousand years' (Ludovic Slimak et al. (eds), *Le Troisième Homme. Préhistoire*

de l'Altaï, exhibition catalogue, Éditions de la Réunion des musées nationaux-Grand-Palais, 2017, p. 130).

5 Ludovic Slimak, *The Naked Neanderthal*, translated by David Watson (London: Penguin, 2023).

6 The Levallois technique of flint knapping is named after the fact that such flints were found in the Levallois–Perret suburb of Paris, where there is a metro station. (Translator's note.)

7 Theodora Kroeber, *Ishi in Two Worlds. A Biography of the Last Wild Indian in North America* (Berkeley, CA: University of California Press, 1961), p. 3.

8 Lucien Bodard, *Green Hell: Massacre of the Brazilian Indians*, translated by Jennifer Monaghan (New York: Outerbridge & Dienstfrey, distributed by E. P. Dutton, 1971).

9 Ibid., pp. 26–27.

2 *The last Neanderthal? An improbable encounter*

1 Philip the Fair (Philippe le Bel) was King of France between 1285 and 1314. (Translator's note.)

2 Jean-Paul Sartre's play *Les Mains sales* (*Dirty Hands*) deals with the difficulty of keeping one's hands clean and one's motives pure in politics. (Translator's note.)

3 The burle is a winter wind that blows from the north across south and central France. (Translator's note.)

4 Aldous Huxley, The Mike Wallace Interview Collection, 18 May 1958 (https://www.youtube.com/watch?v=1ePNGaom3XA). But see also his lectures at various American universities, including Berkeley: https://digital.library.ucla.edu/catalog?utf8=%E2%9C%93&per_page=100&view=list&q=aldous+huxley&search_field=all_fields.

5 Avi Loeb, *Extraterrestrial. The First Sign of Intelligent Life Beyond Earth* (New York: Houghton Mifflin Harcourt, 2021).

3 *How do humans die?*

1 Boris Cyrulnik, *Le Laboureur et les Mangeurs de vent* (Paris: Odile Jacob, 2022).

2 Frank Herbert, *Dune*, in *The Great Dune Trilogy* (London: Gollancz, 1979), p. 16.

3 Henri Bergson, *The Two Sources of Morality and Religion*, translated by R. Ashley Audra and Cloudesley Brereton with the assistance of W. Horsfall Carter (London: Macmillan, 1935), p. 7.

4 Ibid.

5 Allen Madden, 'First contact: a contemporary Aboriginal perspective', *Stories & Ideas*, Museum of Contemporary Art Australia, 2009, available online: https://

www.mca.com.au/about-us/mca-story/first-contact-contemporary-aboriginal
-perspective/#:~:text=Gadigal%20Elder%20Allen%20Madden%20imagines
,typical%20summer%20day%20in%20Warrane.

6　Ibid.

7　Judy Klemesrud, 'The Gods Must Be Crazy – a truly international hit', *The New York Times*, 28 May 1985.

8　See Isak Dinesen (pen-name of Karen Blixen), *Out of Africa* (London: Penguin, 2001).

9　An allusion to Claude Lévi-Strauss's 1955 book *Tristes tropiques,* a study of the 'sadness' of tropical peoples following their often catastrophic encounter with Westerners. See Claude Lévi-Strauss, *Tristes tropiques,* translated by John and Doreen Weightman (New York: Atheneum, 1973). (Translator's note.)

10　During my time in the US and Canada, I frequently encountered the disturbing notion – expressed by individuals, ranging from everyday people to researchers and museum directors – that Native Americans no longer genuinely held their ancestral traditions, dismissing these as modern constructs or inventions. This short passage challenges that perspective by arguing that if Native American traditions are considered reconstructed and thus illegitimate, then the same must apply to White traditions. Traditions inevitably fade with those who uphold them; all humans continually reinvent their identities, generation after generation, in a perpetual state of cultural amnesia. Rather than judging, this reflection seeks to dismantle condescending attitudes and reposition cultural loss and reinvention as universal human conditions.

11　The last lines of T. S. Eliot's poem 'The Hollow Men'. (Translator's note.)

12　See Arnold Toynbee, *Civilization on Trial* (Oxford: Oxford University Press, 1948) and Jared Diamond, *Collapse: How Societies Choose to Fail or Succeed,* 2nd edn (London: Viking, 2011).

Index